IMAGES
of America

NEW JERSEY'S COVERED BRIDGES

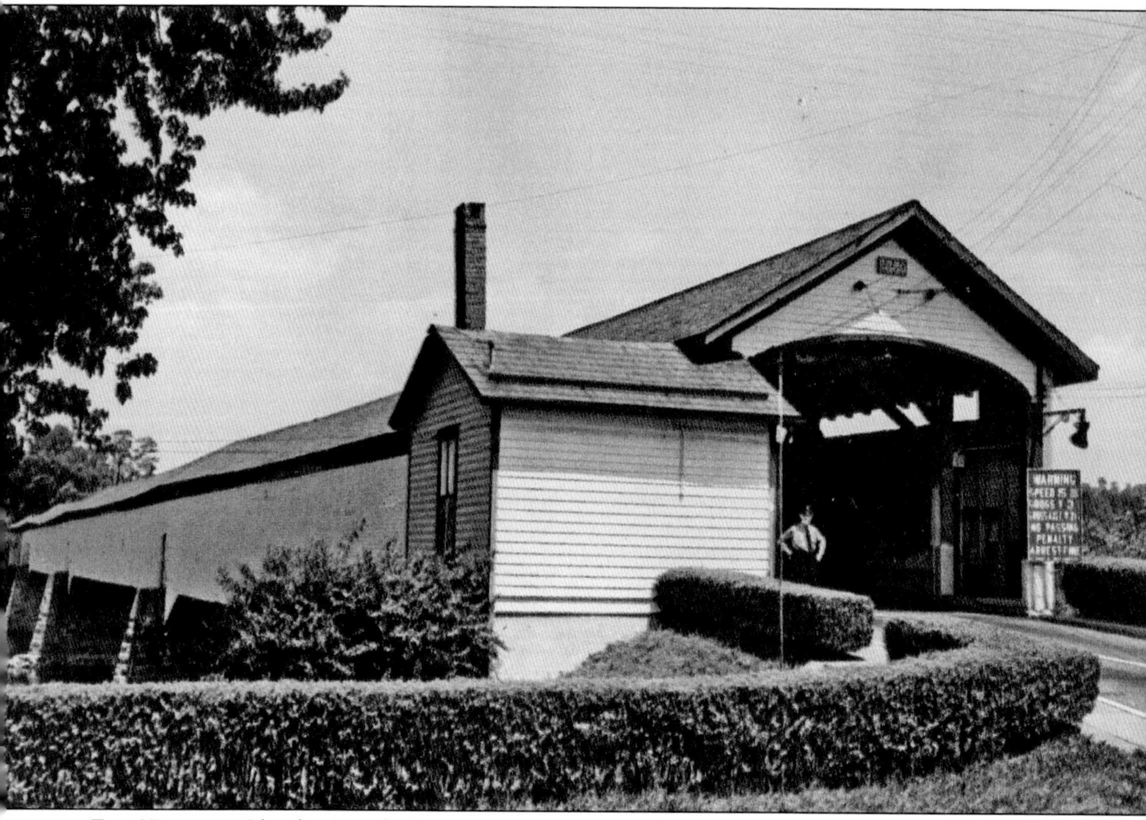

For 37 years, Charles Newbaker Sr., pictured, worked at the Portland-Columbia Covered Bridge. When tolls were eliminated in 1927, he continued as a guard, chief carpenter, historian, and protector. He was the last man to walk across the bridge before it was destroyed in 1955. (Author's collection.)

On the Cover: The Portland-Columbia Covered Bridge was the last of the dignified and strong line of wooden covered bridges that crossed the Delaware River between New Jersey and Pennsylvania. Weakened by age and nature, the bridge built in 1869 had been designated as a pedestrian-only passage in 1953, as seen here. (Author's collection.)

IMAGES
of America

NEW JERSEY'S COVERED BRIDGES

Richard J. Garlipp Jr.

ARCADIA
PUBLISHING

ISBN 978-1-4671-2011-1

Published by Arcadia Publishing
Charleston, South Carolina

Printed in the United States of America

Library of Congress Control Number: 2013931616

For all general information, please contact Arcadia Publishing:
Telephone 843-853-2070
Fax 843-853-0044
E-mail sales@arcadiapublishing.com
For customer service and orders:
Toll-Free 1-888-313-2665

Visit us on the Internet at www.arcadiapublishing.com

To all the great folks whom I have met along my journey: may you continue to pursue, to discover, and to enjoy our New Jersey history!

Contents

ACKNOWLEDGMENTS

Many, many thanks to many people: my lovely wife, Patti, and my great sons, Rick and Matt, for both moral and technical support; my sister, Pat Garlipp; David Wright, president of the National Society for the Preservation of Covered Bridges (NSPCB); Bill and Jenn Caswell, NSPCB; Bob and Trish Kane, NSPCB and Burr Resource Center, who suggested this project to me; Linda Schwart, Broadway Printers, Long Branch; Thomas Kipphorn, Todd Clark, and William Umek, for invaluable information and images; Betsy Dudas and Ferdinand Klebold, Ocean County Historical Society; Wini Smart; Bob Mitchell; Wendy Nardi and Allison Tune, Trentoniana Room; Don Cornelius, Hunterdon County Historical Society; Sue Dolbow, Salem County Historical Society; Marge Dombrosky, Swedesboro, Gloucester County Library System; Dr. Bob Belvin, New Brunswick Free Public Library; Joe Donnelly, Delaware River Joint Toll Bridge Commission; Barbara Price, Cloucester County Historical Society; the late John Adams; Jack Connolly, Dennisville Historic Home Owners Association; Bob David; Jeff McVey, Lambertville Historical Society; Sam Pharo, Kinnelon Public Library; Stephanie Gabelmann, Boonton Public Library; Carol Phillips; Mildred Wehr; Melinda Taylor; Bill Turkowski; Tom Glover; Joanne Nestor; Dr. James Amemasor and Doug Oxenhorn, New Jersey Historical Society; Nancy Polhamus, Gloucester County Library System; Alicia Batko, Richard and Beth Jones, Montague Association for Restoration of Community History; Carol Elfo; Nancy O'Brien; Denise Shelton Trocchia; Peter Joseph; Ed and Wendie Fitzgerald; Marfy Goodspeed. One final thank-you to my persistent yet patient editor at Arcadia, Katie McAlpin. Unless otherwise noted, all photographs are from the author's collection. The United States Department of Transportation Federal Highway Authority (USDOT-FHA) provided wooden bridge structural designs.

INTRODUCTION

At some point in time, travelers in many areas of the United States may glimpse an unfamiliar wooden structure standing alone alongside the highway. At first sight, it is probably thought to be a barn, but upon further review, it appears to be a barnlike structure crossing a stream or small river like a bridge. A bridge. A covered bridge! Then, with the speed of the highway, it vanishes into the distance.

Often, this quick encounter with history is forgotten or mentioned as a curiosity as the traveler recounts the more elaborate points of interest observed or visited on the trip. Occasionally, however, this moment instills a question that requires more information. Asking a local resident about that unusual wooden structure, or perhaps even researching the unfamiliar term "covered bridge" may bring some sense of awareness and satisfaction.

Speaking with a local resident normally reveals the great pride a town possesses for its covered bridge. And use of the Internet quickly indicates the large number of covered bridges that existed, the important historic roles they played, and the widespread disappearance of these structures.

All of this is especially true within the state of New Jersey, unlike neighboring or not-so-distant states like Pennsylvania, New York, Vermont, and New Hampshire, which all today still boast a fair number of covered bridges to visit and appreciate. Even though many have vanished, New Jersey can direct visitors to one extant covered bridge!

The Green Sergeant's Covered Bridge, crossing the Wickecheoke Creek near Flemington and three miles from the Delaware River, is the subject of a story so often repeated throughout the counties and towns of New Jersey.

Built in 1872 (or 1867, by some accounts), this bridge took the Sergeantville-Rosemont road across the creek in a picturesque valley. The original bridge at the site, named for the son of the builder, Charles Sergeant, was washed away in a flood in the mid-19th century and replaced by the current bridge. Successful in its role for many years, the bridge was badly damaged when an overweight truck attempted to cross it. The headline in New York's *Herald Tribune* of January 15, 1960, read, "Last of covered bridges in Jersey shut to traffic." Only the determination of the Green Sergeant's Covered Bridge Society, a community grassroots group of local citizens, saved the structure. The bridge was dismantled and rebuilt with steel supports below to allow greater loads to cross. Next to the restored covered bridge, an open concrete bridge was built to handle one direction of traffic. A stone marker near the bridge tells its story: "In 1961, as a result of the efforts of an aroused group of citizens, the State of New Jersey, using the materials of the original covered bridge, fully restored this link with the past."

Briefly, this is the history of one covered bridge in the state of New Jersey, but the story was repeated again and again throughout the regions of the state. Unknown or little-known names like Dennisville, Dividing Creek, Crosswicks Creek, Three Bridges, Salem, and Tuckahoe were unceremoniously replaced by iron, steel, and concrete, but with no effort to bypass and preserve the original structures.

Or nature can wreak havoc before the progress of humankind even enters the equation. An example in point was the beautiful Centre Bridge crossing the Delaware River at Stockton, New Jersey. Opened in 1814, this toll bridge was located between the two existing covered bridges up- and downriver. With some repairs and redesign work being done early on, the bridge served the people well. But then it was nature's turn. An ice-choked river flood in 1841 took out three of the six spans of the bridge. Repairs immediately followed, and the bridge was raised six feet higher over the Delaware. A major flood in 1862 did not cause serious damage, and the bridge remained open. Again in 1903, Centre Bridge survived a huge ice flood that damaged or destroyed river crossings between Phillipsburg and Trenton. Only two years later, a major fire in Stockton, which burned many structures, was stopped right at the bridge portal. Some two decades later, another fire inflicted some damage but was repaired. Then nature finally wrapped up its assault. One evening in the summer of 1923, lightning struck the bridge, and the resulting fire completely destroyed Centre Bridge. Only the stone piers remained, and the bridge was never rebuilt until 1927, then owned by the Delaware River Joint Commission. Ferries were used again for crossing.

Forests were abundant in the early years of the state. Roads began to be planned out, often using the original Native American trails. Creeks, streams, and rivers needed to be crossed. Simple stringer bridges were the first structures to allow quick and safe passage. Ferry companies were chartered to cross larger rivers, but with the increase in population and commerce, ferries became time-consuming and cumbersome. The early stringer bridges could handle only so much weight and volume. Longer, wider, and stronger bridges became necessary. This ushered in the era of the major bridge builders and the truss designs to accomplish this task.

In Hunterdon County, the freeholders began to levy taxes around the turn of the 19th century to build bridges over inland creeks and rivers. Road conditions, however, were very poor, and some companies were incorporated to build turnpikes and charge tolls that would be used for maintenance of these roads in all seasons, a provision which sometimes proved successful. But covered bridges were built not only in Hunterdon but also throughout the state. The actual number varies across different accounts, from 50 to 75, although 75 would be a difficult tally to authenticate. As significant as their contributions were to the history of bridge technology in New Jersey, covered bridges were not without inherent weaknesses that eventually made them defunct as stiffer, stronger bridge types requiring much less maintenance were developed.

But New Jersey did have its covered bridges, and this little book will attempt to present and describe the photographs that illustrate this history. Chapter introductions will describe historically authenticated covered bridges for which no image has been found. Captions will describe the images accurately. Every effort has been made to offer a work as complete and interesting as possible.

New Jersey is a wonderful place, full of life, diversity, and history. Please enjoy this particular aspect of its history: *New Jersey's Covered Bridges*.

One

The Truss and the Roof

Andrea Palladio, a Venetian architect, is usually credited as the first to describe the structural form recognized as a truss. France became the leader in early bridge engineering, but it was based on stone and arch theory. The Swiss and the Germans devoted their attention to using timber trusses in their bridges. Other developments in the evolution of timber-truss bridges followed in several other European countries, but early bridge building in the United States led to the most significant advancements in the theory of truss behavior.

A notable advancement in timber bridge building was the crossing of the Connecticut River at Bellows Falls, Vermont. Col. Enoch Hale used a two-span structure with a total length of 365 feet. The supporting structure was a strutted beam.

Another American bridge pioneer was Timothy Palmer, who made consistent use of more panel-braced timber frames in configurations that can be identified as trusses. Palmer gained approval to span the Schuylkill River in Philadelphia, Pennsylvania. Completed by 1805, his bridge included three spans, with the trusses built of heavy timber members and bracing. It was enclosed with sides and a roof to protect it from weathering. Although there are hints of even earlier covered bridges in the United States, this bridge is most often cited as the first.

Early North American bridge builders actively pursued patents for their designs to gain more contracts. Timothy Palmer received a patent in 1797. Theodore Burr obtained the first of his many patents sometime between 1804 and 1806, while the patent for his trademark design was issued in 1817. Lewis Wernwag obtained a patent for his 340-foot trussed-arch span, which became known as the Colossus. Ithiel Town's simple lattice-style bridge was patented in 1820, and Stephen Long received his first bridge patent in 1830. William Howe was the first to use metal components, namely iron rods, as the primary components within an otherwise timber truss, which could withstand the heavy loads of the railroad.

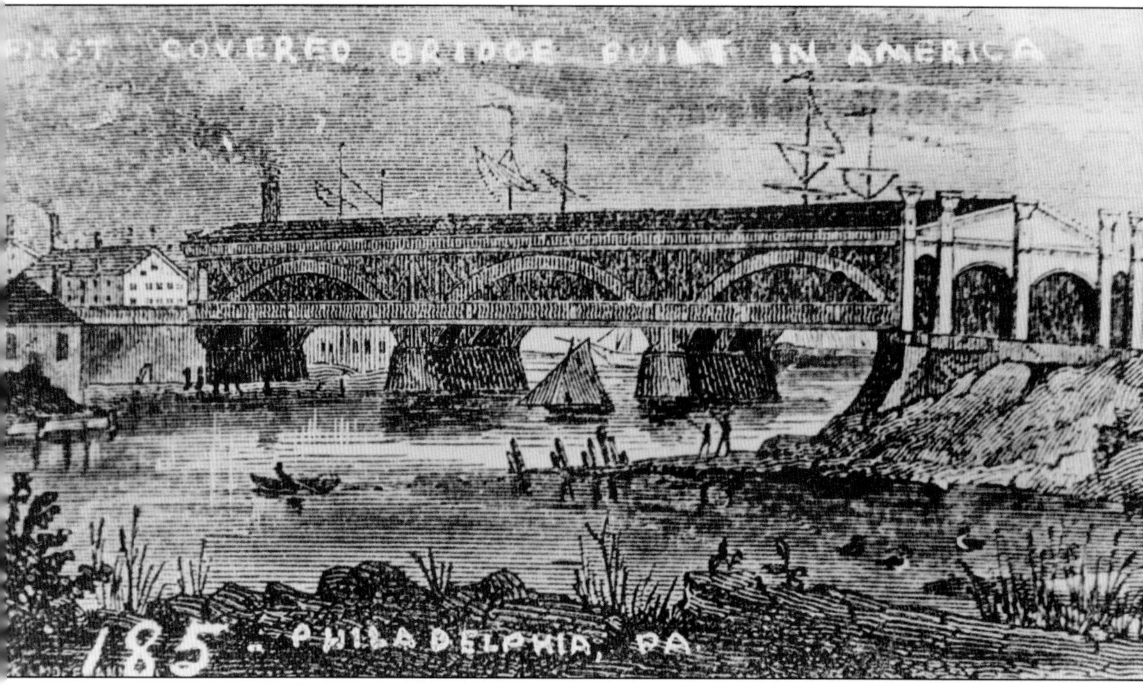

Timothy Palmer built the first covered bridge in the United States, seen above, over the Schuylkill River in Philadelphia. Palmer was an energetic and prolific builder who experimented with progressively flatter structures that relied less on arch action. The bridge was expensive and critical to ongoing commerce. It became known as the Permanent Bridge. Another early covered bridge, the Colossus, was designed and built by Lewis Wernwag. The huge span represented a major triumph in bridge construction, with its attractive and apparently efficient use of timber supplemented with iron rod bracing members. Wernwag owned a metalworks company and relied more on early forms of metal connections and components rather than only traditional timber joinery.

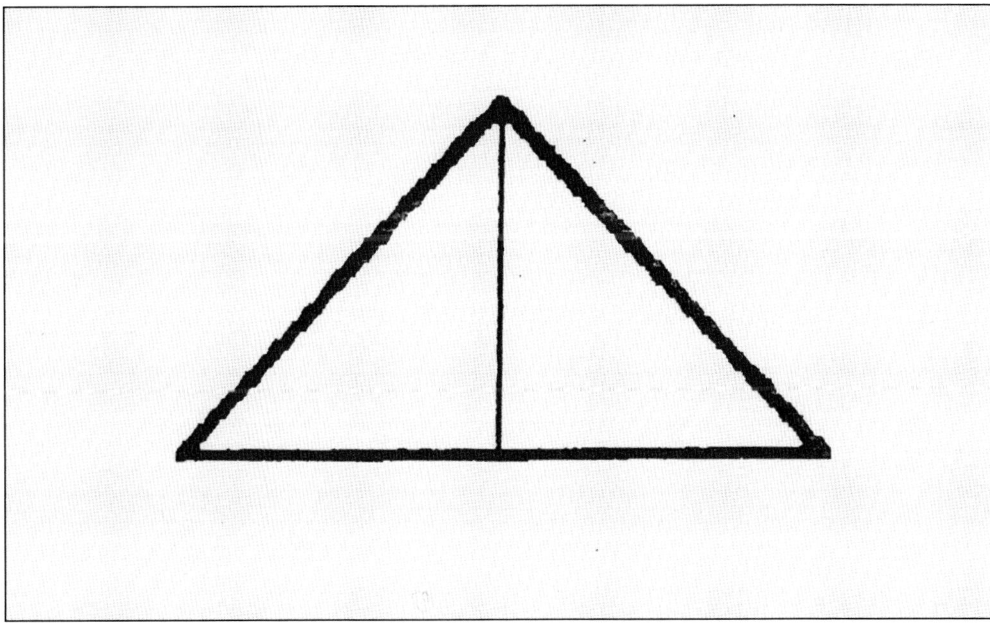

The most elementary, heavy timber-truss configuration is the kingpost. The inclined members of a kingpost truss serve both as the top chord and as the main diagonals and resist compression forces. The horizontal member along the bottom of the truss is the bottom chord, which acts in tension. The central vertical kingpost acts in tension to support the floor load. (Courtesy of USDOT-FHA.)

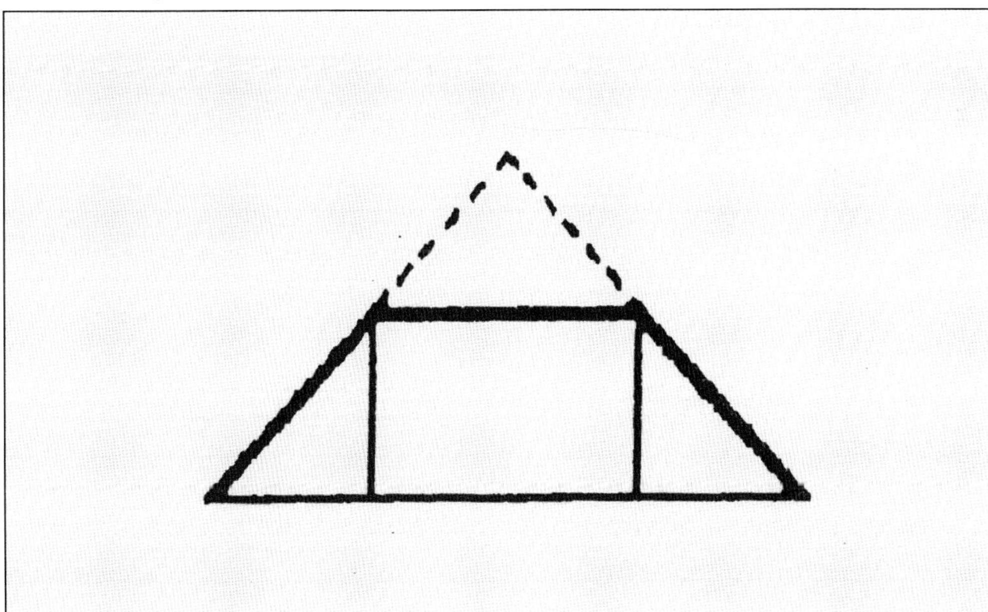

The next range in span lengths commonly includes trusses developed from a simple modification of the kingpost. The queenpost truss is, conceptually, simple: a stretched out version of the kingpost truss accomplished by adding a central panel with extra horizontal top and bottom chords. Queenpost-truss bridges range from about 40 to 60 feet, with a few known to have been longer. (Courtesy of USDOT-FHA.)

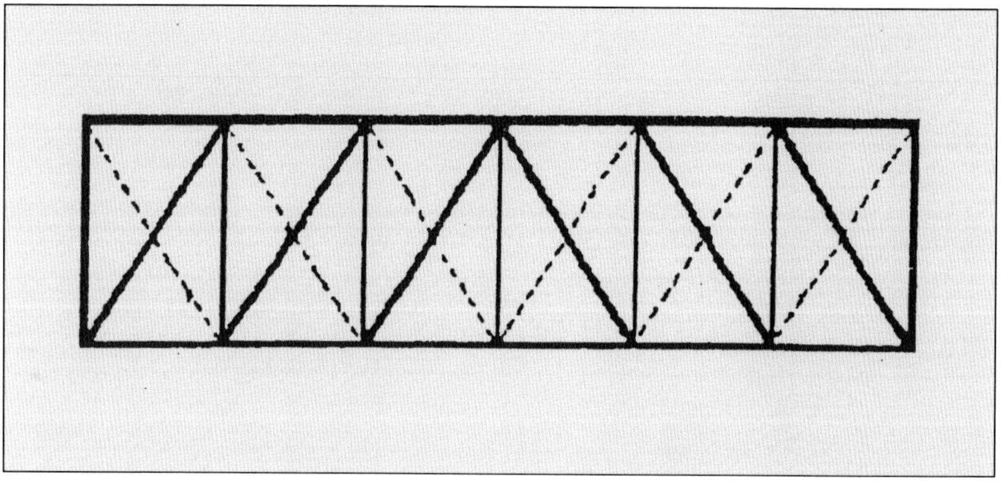

A straightforward way to stretch the span capability of the queenpost truss is adding panels to the kingpost truss to create what is known as multiple-kingpost trusses. Most of these trusses were built with an even number of panels so that all the diagonals were in compression and all the verticals were in tension, under normal loading. (Courtesy of USDOT-FHA.)

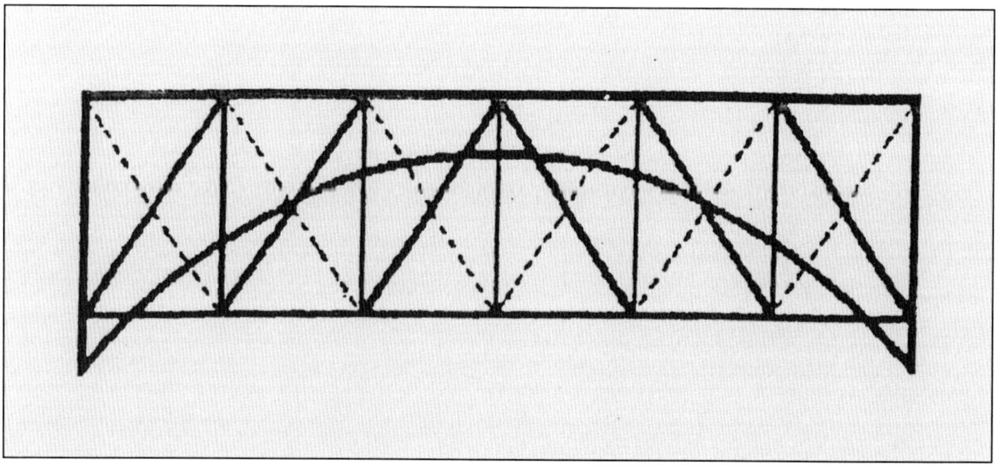

The Burr arch is basically a combination of a typical multiple-kingpost truss with a superimposed arch. The arch was added to allow for heavier loads on the bridges and to stretch their span capabilities to greater lengths. Surviving examples of Burr-arch bridges have spans up to 222 feet. The actual arches are in pairs and sandwich a single multiple-kingpost truss. (Courtesy of USDOT-FHA.)

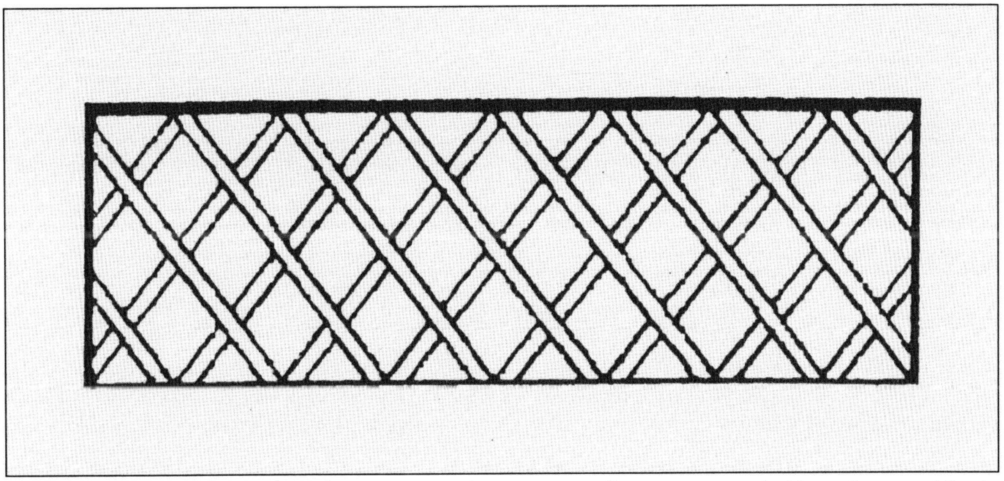

Ithiel Town, an architect by education, sought a means of constructing bridges that would rely on an easily adapted design and would require less skilled labor. The design could extend to a wide range of span lengths. In the opinion of many, Town's patented truss represents the most important development in the history of covered bridges, with a popular and enduring style. (Courtesy of USDOT-FHA.)

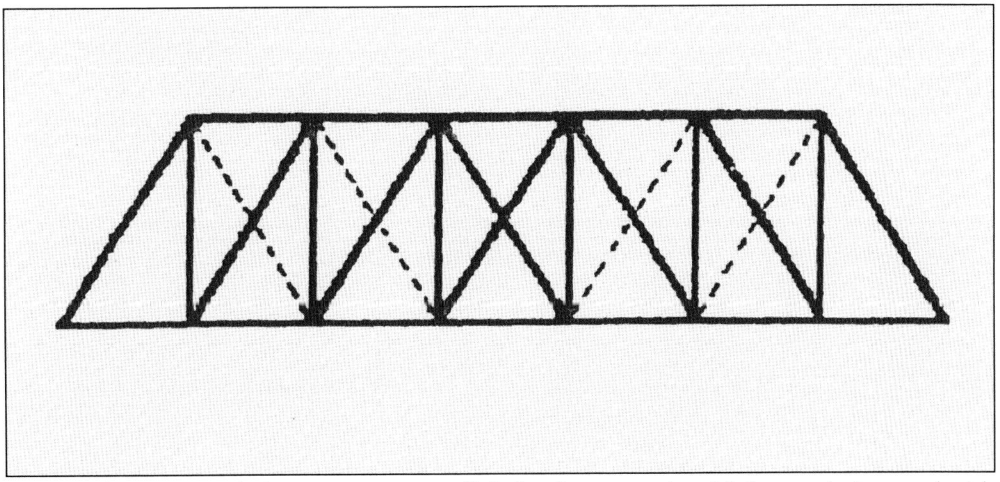

Col. Stephen H. Long's focus was on a parallel chord truss made with heavy timbers and with crossed diagonals in each panel. A special feature of his bridge included the use of timber wedges at the intersection of the chords, posts, and diagonals. The Long truss was adopted by many builders for use in highway and railroad bridges. (Courtesy of USDOT-FHA.)

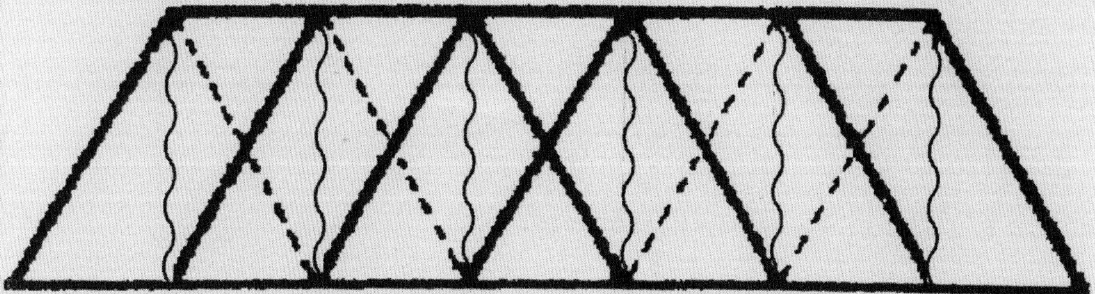

William Howe of Massachusetts was granted his first truss patent in 1840 and a second one later in the same year. His second patent used straight metal rods as the vertical members of what was otherwise a simple timber parallel-chord, cross-braced truss. This was the first truss patent granted with major structural components made of metal. Little skilled labor was involved in assembling and erecting this type of truss, and it became an immediate success. Another factor in the success of Howe's truss type was his inclusion of a detailed structural analysis with the patent application. The Howe truss is second only to the Burr arch in popularity of extant covered bridges in the United States. There are many other truss configurations that were patented, including those identified as Smith, Paddleford, Pratt, Childs, and Partridge trusses. The Pratt truss deserves special note because it was the precursor of the popular metal truss of this type. Very few Pratt timber-truss bridges remain, but a very large number of Pratt metal trusses survive. (Courtesy of USDOT-FHA.)

Two

The Delaware River Region

This area of the state of New Jersey includes Burlington, Camden, Gloucester, Mercer, and Salem Counties. For the most part rural and very much the source of New Jersey's moniker as the "Garden State," the Delaware River region is home to Jersey tomatoes and large areas of cranberry bogs and blueberries thriving in the sandy soil. The state capital, Trenton, is located in Mercer County, where two amazing covered bridges crossing the Delaware River were located in the 19th century.

Ferries were the prominent means to cross the rivers and large creeks in the 18th century. But the time factor and load restraints brought the need for bridges to the forefront. According to *Road Records of Gloucester County*, an application was submitted to develop a road running to the "new bridge erected from Samuel Spicer's Landing across Cooper Creek." Upstream near Haddonfield, the Cooper River is known as Cooper's Creek, named after the Cooper family who were among some of the first European settlers in the area of Camden County. This document was dated March 8, 1762, when the new bridge was completed. Becoming known as both Spicer's Landing and Cooper's Creek, the bridge was kept in good repair and rebuilt as a truss bridge in 1833.

In Salem County, Revolutionary War activity recorded in the *Historical Collections of the State of New Jersey* reveals the name of the Hancock Bridge at the marshes near the mouth of Aloes (now Alloways) Creek, but no public road led to this bridge. In 1778, the British, under a Captain Dunlop, seized this bridge and raided the house of John Hancock, killing over 20 "rebels." A footway lead to the second bridge, Quinton's Bridge, and the British Queen's Rangers pursued the colonial militiamen there. Only one escaped this attack on March 18, 1778, known as the Battle of Quinton's Bridge.

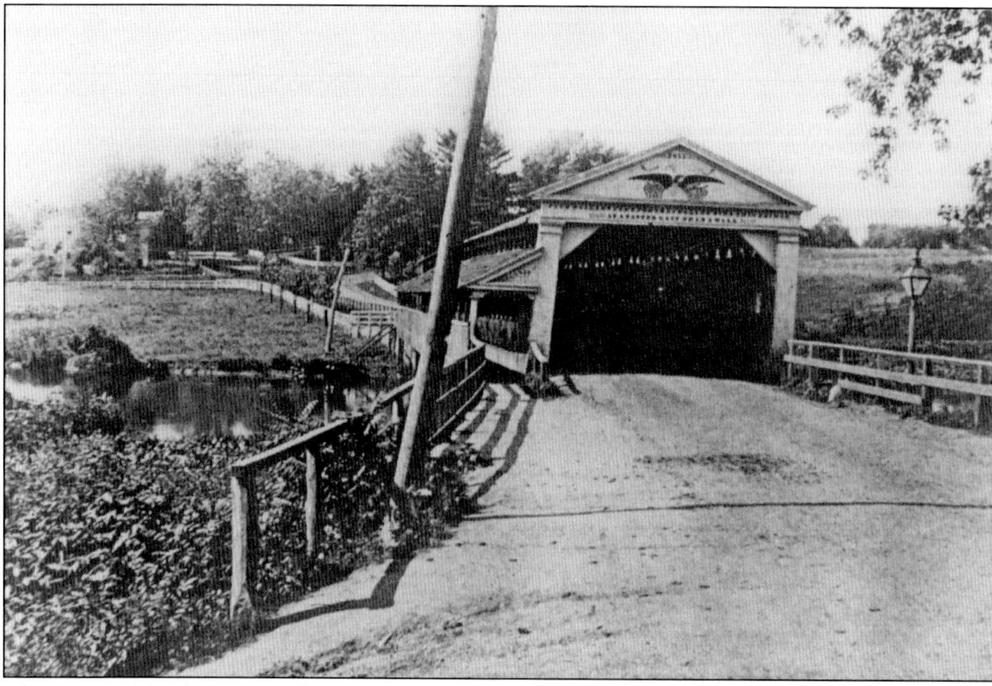

This covered bridge spanned the Crosswicks Creek between Burlington and Mercer Counties on Church Street from 1833 to 1908. In 1906, the freeholders met to discuss demolishing the covered bridge. It was sold at auction in 1908 and replaced by a new iron bridge. A barn in North Crosswicks was built from some of the bridge's old timbers.

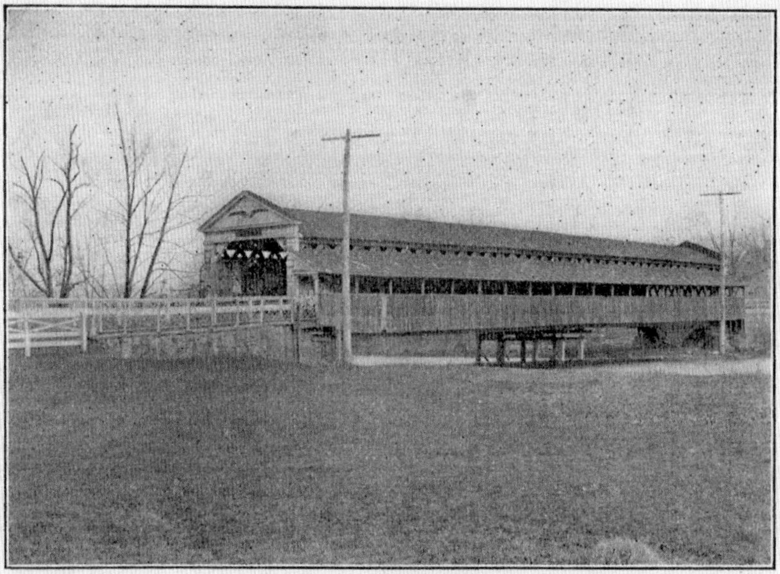

No. 26—Old Bridge over Crosswicks Creek, Crosswicks, N. J.

The Crosswicks Creek Bridge was described as being built in the old-style lattice truss, without a nail being driven into it. Henry J. Bazzel painted American eagles on both portals in 1856, and a walkway was added to the west side of the bridge, which was enclosed in glass.

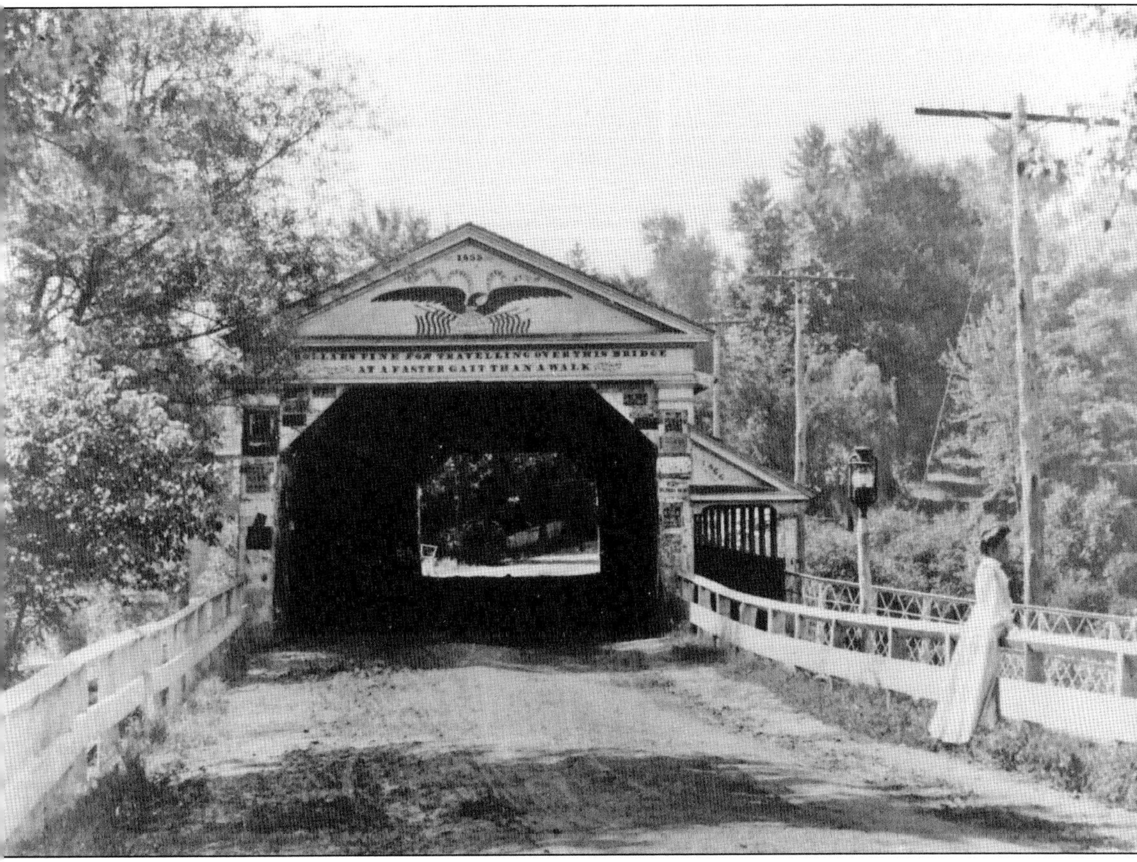

This captivating image shows Jennie Woolman Scattergood looking over the approach rail. In more turbulent times, the Hessians crossed the first bridge at this site during the Revolutionary War, before the Battle of Trenton. Without a ceremony, a new bridge between North Crosswicks and Crosswicks proper, this time made of iron, was opened to the public on December 1, 1908, attended only by Burlington County Board of Freeholders employees. No one was present to make the first trip over. Crosswicks Creek is the dividing line between Mercer and Burlington Counties, and both governments had to share the expense. The first bridge was built sometime around 1738, and the lattice-truss covered bridge seen here replaced it in 1833, lasting 75 years. (Courtesy of Thomas Glover.)

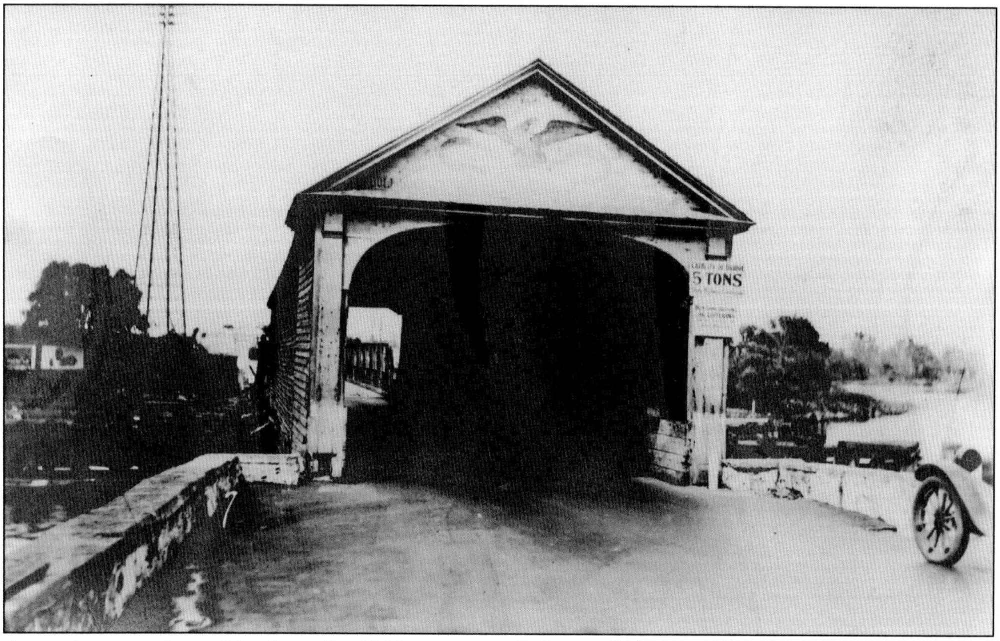

A north-to-south road built in the 1740s connected the settled town of Burlington with Cooper's Ferry (later Camden). A ferry began operation in 1749 to cross Rancocas Creek. The new road and ferry laid the groundwork for business in the area and for the eventual growth of the town of Bridgeboro, Burlington County. This town obviously was given its name at a later interval, after the first bridge over the Rancocas succeeded the earlier ferry. Historian George Decou gives the date of the first bridge, a covered one, as 1793. In 1838, a new bridge called a double draw, a covered Burr arch, succeeded it. The portal of this bridge displayed a large painted eagle.

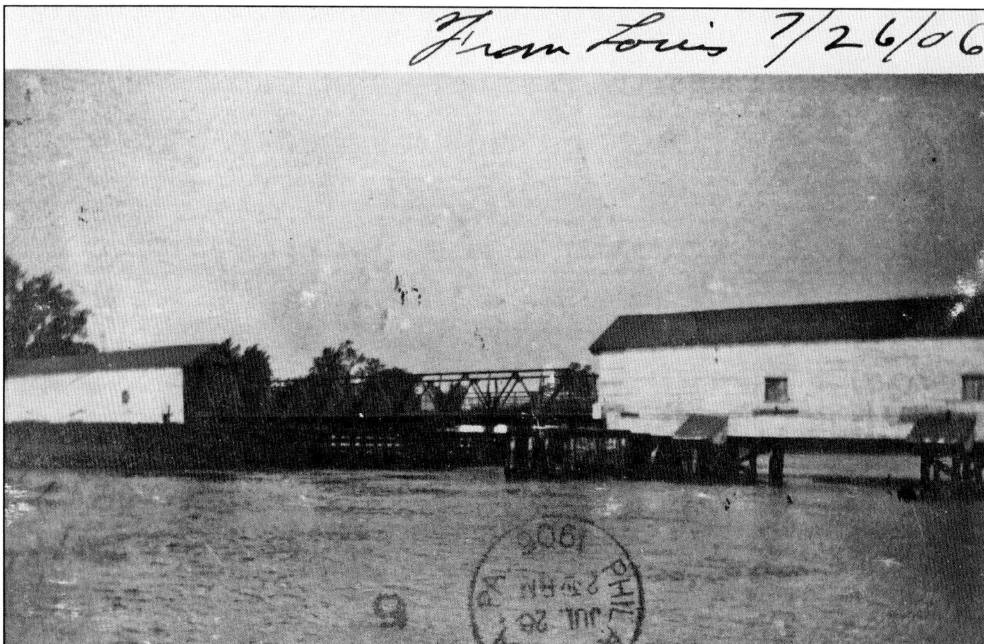

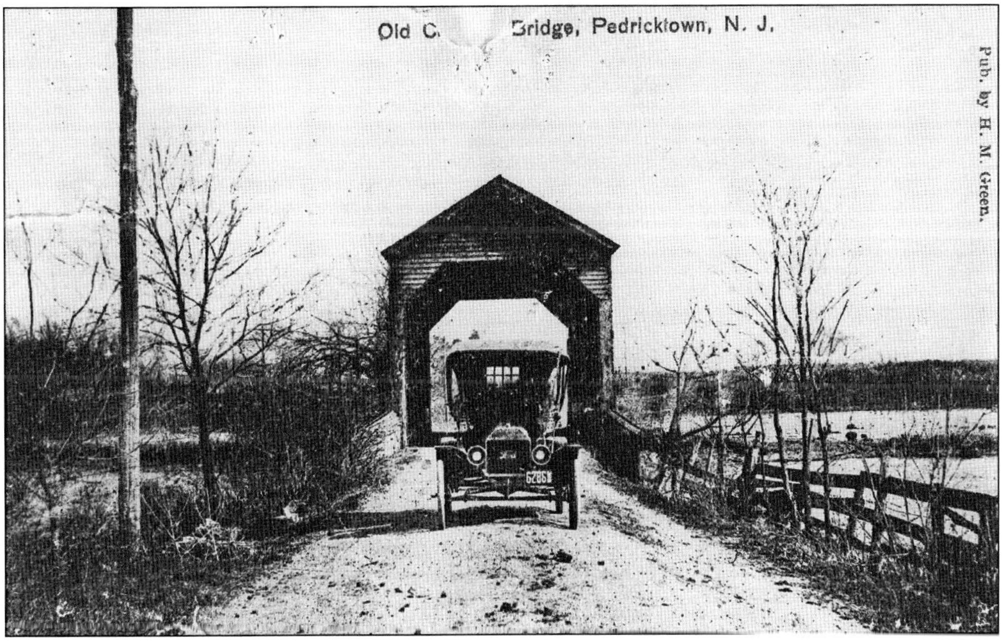

Covered bridges in this area came into popularity around 1805. This covered bridge over Beaver Creek on Auburn Road, which went from Pedricktown to Auburn, existed as late as 1900, when it was reportedly sold to William B. Darlington Sr., who made it into a wagon shed. Salem County reconstructed this bridge area in 1921. Purchasing 1,000 acres of John Fenwick's land, Roger Pedrick and his descendants were the first to establish a permanent settlement in 1674, called Pedricksburg. By 1883, a steam mill on Oldmans Creek was the largest business in Pedricktown. In 1880, George R. Pedrick established a seed business, which was one of the largest distributors in the United States.

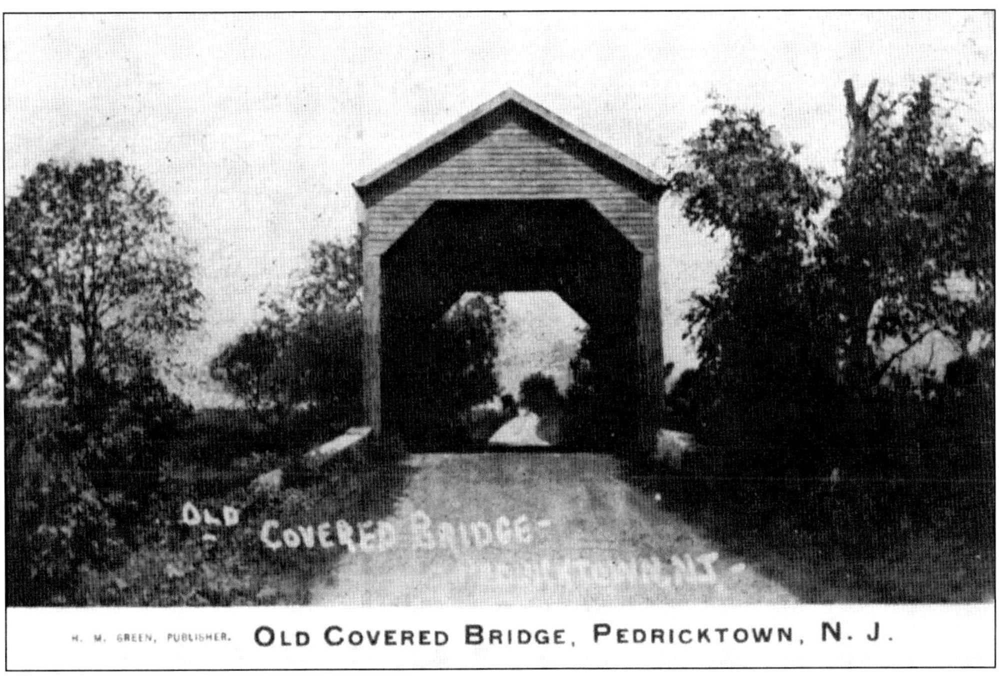

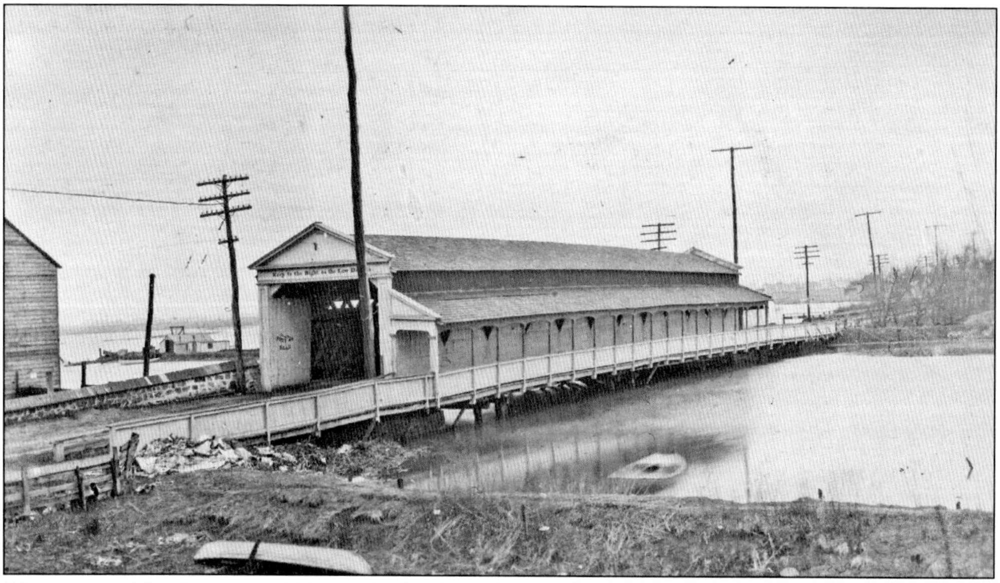

This covered bridge crossed Fenwick Creek in the southern New Jersey town of Salem. Built by Amos Campbell in 1831, the single-span bridge was of Town lattice-truss design, built close to the water since the creek was not a navigable waterway for shipping. A covered walkway was attached to one side.

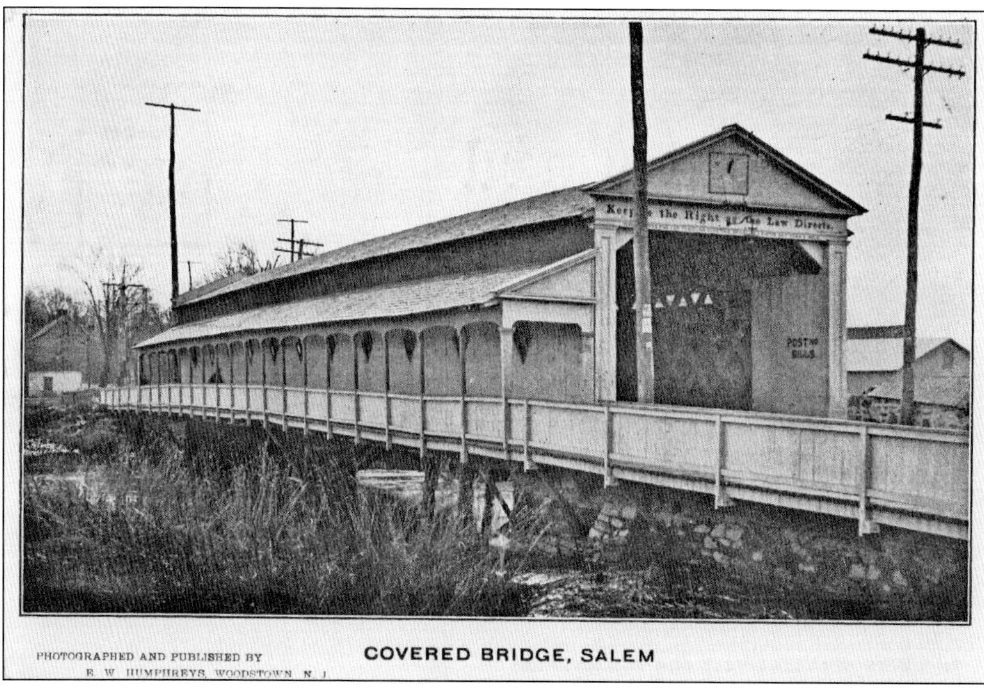

A written comment from the local newspaper *National Standard and Salem County Advertiser* regarding the Salem covered bridge stated, "This is a good bridge today [1887] and if well cared for will last a hundred years. But if the authorities do not keep it lighted at night the demand will soon grow stronger for an open iron bridge in its place." But not until 1922 did an iron bridge replace it.

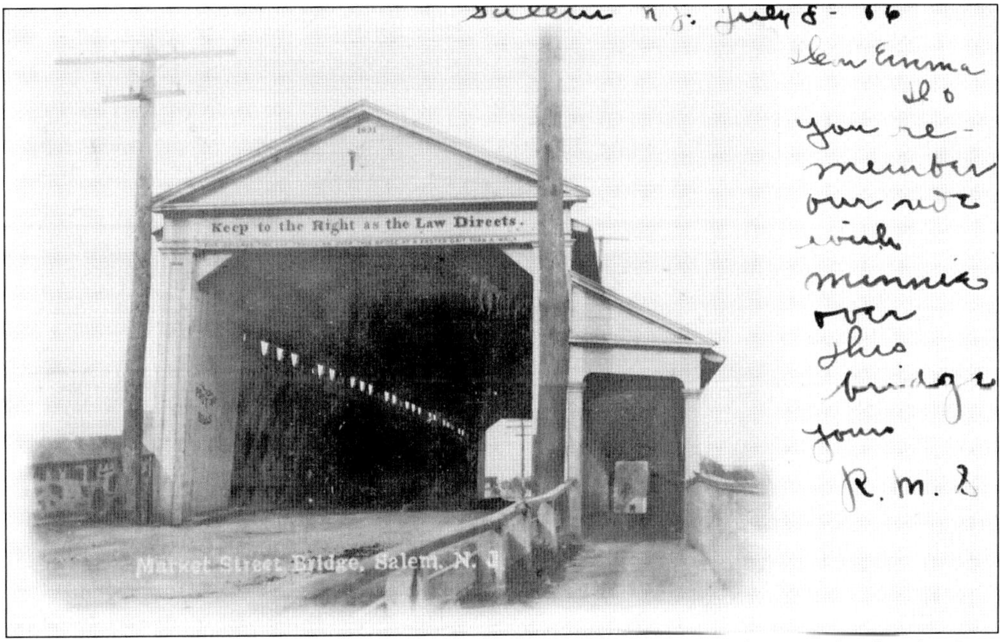

Entering the covered bridge at Salem over Fenwick Creek, the sign on the portal ordered, "Keep to the Right as the Law Directs." Above the directive was a clock, centered in a nicely peaked and extended front piece. The diamond-shaped windows, coordinated with the Town lattice within, provided light for the driver. The bridge was built close to the water level of the creek, and the walkway for pedestrians was covered by a separate roof and secured on the waterside by a high closed railing. The bridge connected Woodstock Road in Mannington Township to Market Street in the thriving town of Salem, seen below. (Above, courtesy of Todd Clark; below, courtesy of the New Jersey Historical Society.)

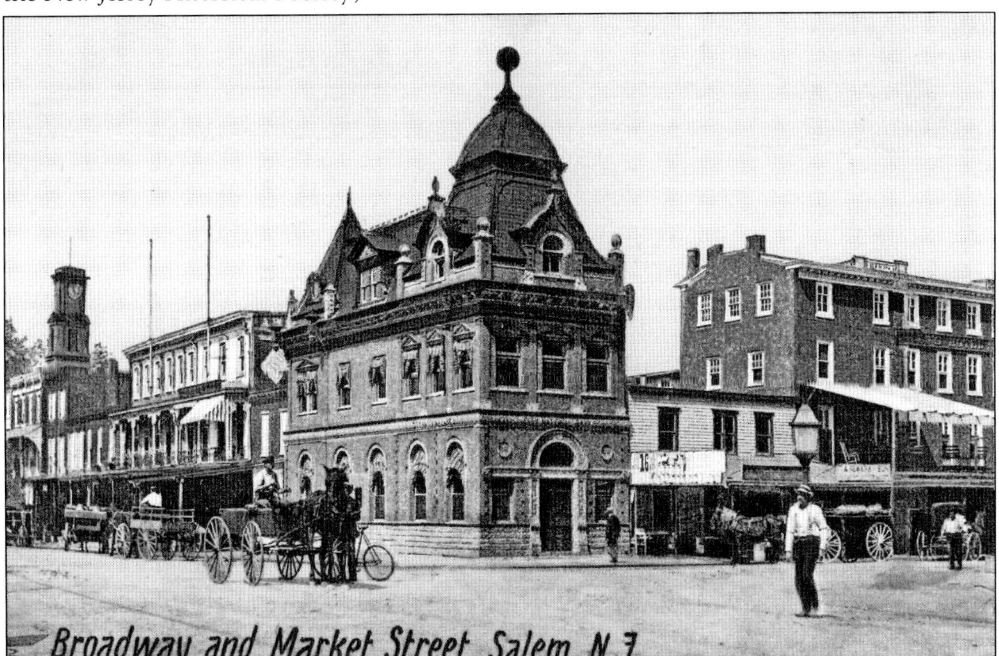

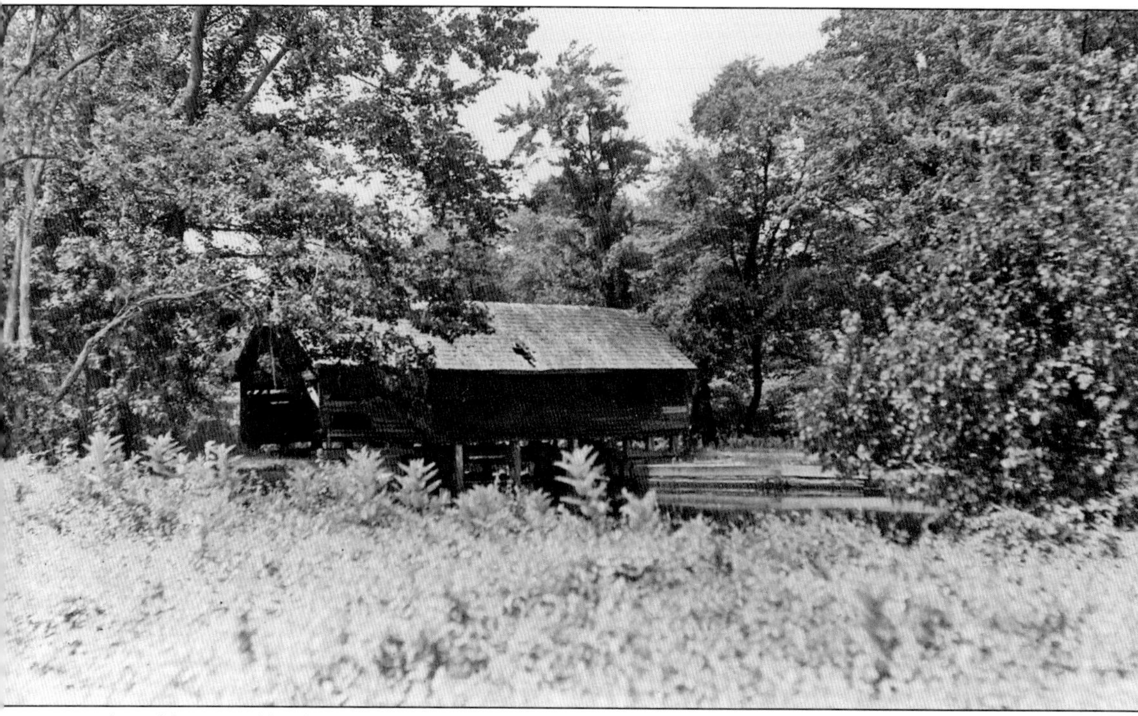

This old covered bridge with floodgates is located in South Pemberton, Burlington County, over the Rancocas Creek. Built in the 19th century, it had a length of 35 feet 5 inches, a kingpost-truss structure, and four spans due to the floodgates beneath. A thriving industrial region in the early 19th century, Pemberton had craftsmen producing clocks and chairs, a cotton manufacturer, a hat factory, and even three shoemakers. But the depression following the Panic of 1837 caused Pemberton's economy to change to the production and processing of agricultural products, such as blueberries and cranberries. The borough was situated between fertile farmland to the west and the Pine Barrens to the east, with a good transportation system. The bridge's location and its floodgates below most likely indicated its use to help with the harvesting of cranberries; the bogs had to be flooded so the cranberries could be raked to the surface and then gathered.

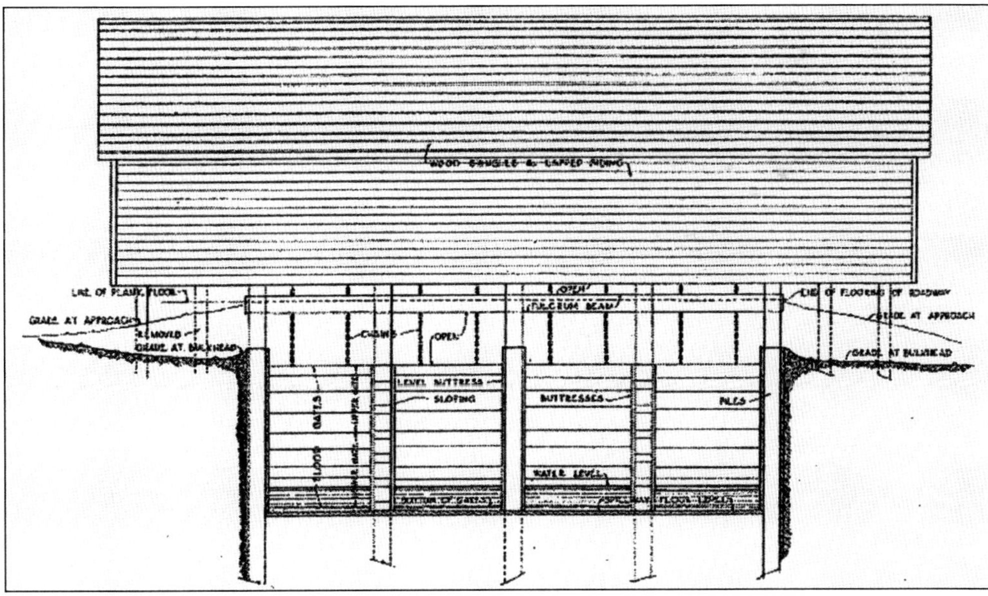

These two diagrams are part of a report from *Covered Bridges Recorded by Historic American Building Survey* of the National Park Service, Library of Congress, Washington, DC. The 1936 report titled "Old Covered Bridge and Flood Gates" includes five pages of drawings with plans, elevations, sections, and details. The above diagram shows the west elevation and the four floodgates to be raised or lowered according to need. The diagram below illustrates the end elevation and details of the windlass. According to the accompanying report, the octagonal, barreled windlass rotates by hand levers to control the upper sections of the floodgates. (Both, courtesy of HABS, NJ, 654.)

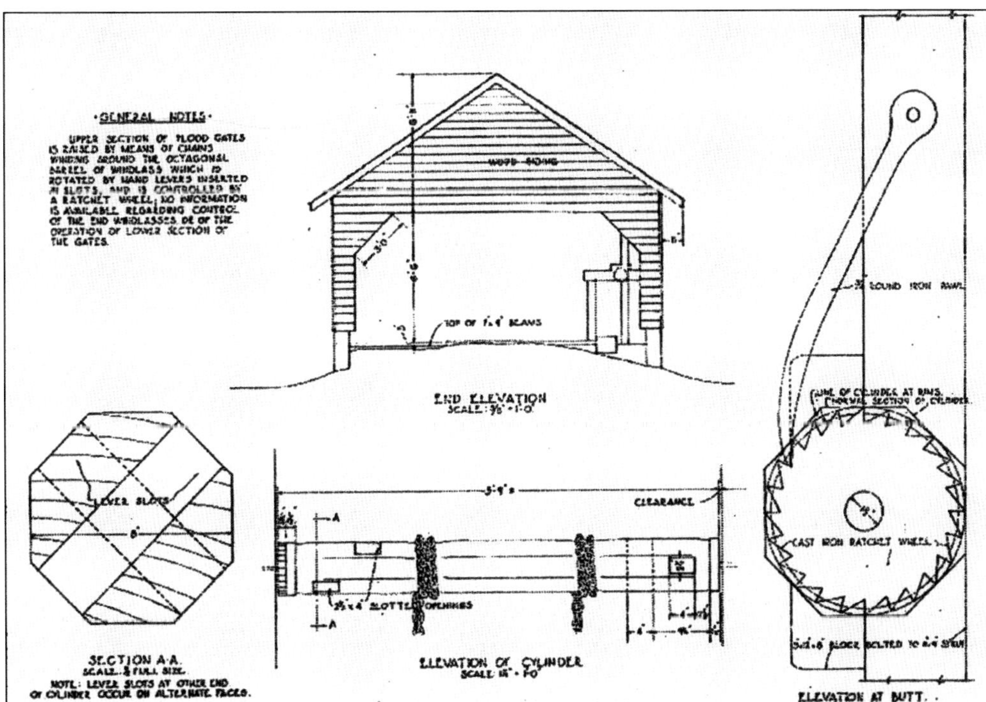

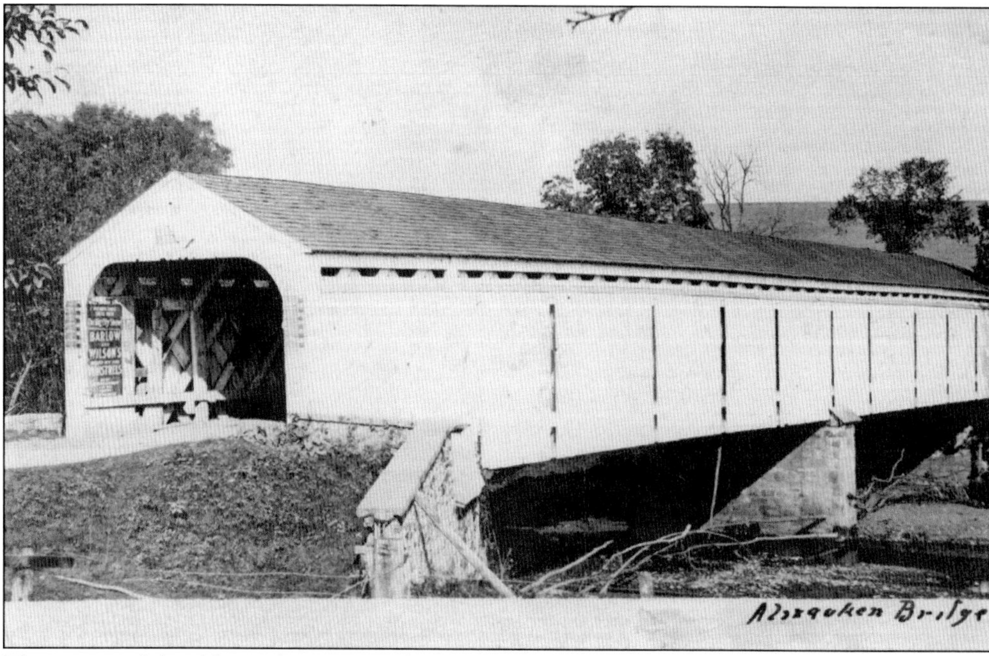

This image of the Alexsauken Covered Bridge, a Town lattice structure, was obtained from the National Society for the Preservation of Covered Bridges. Alexsauken Creek starts in the Sourland Mountain ridge and runs through Hunterdon County. In 1895, a poor English immigrant was found living in the dirt bank of the creek near the bridge. (Courtesy of the NSPCB archives.)

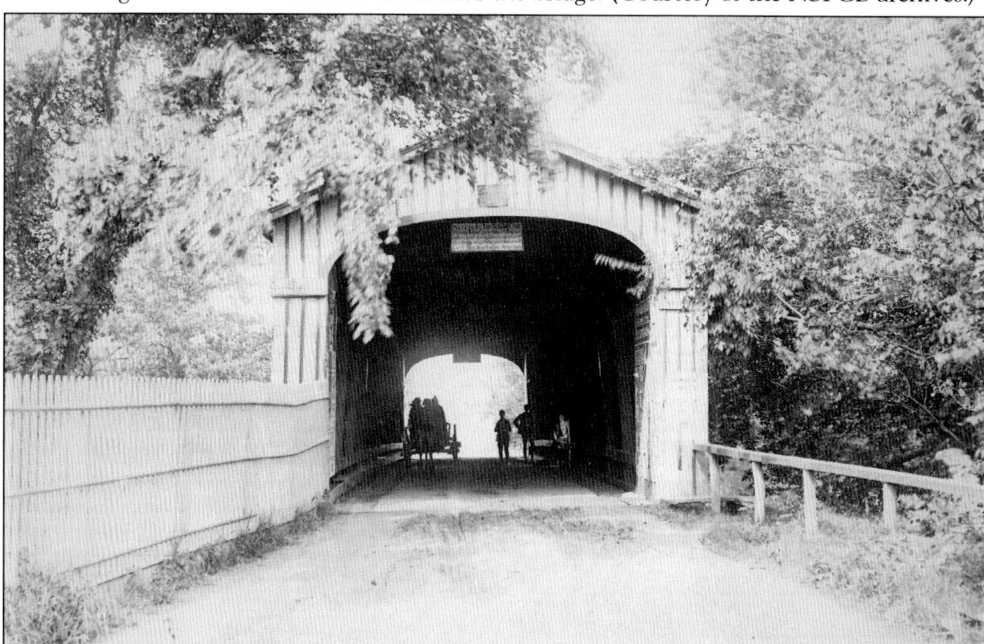

Another archival image from the NSPCB shows the Little Falls Covered Bridge. Passaic County was notable for several early timber-truss covered bridges. This single-span bridge, appearing to have been constructed of multiple-kingpost trusses, was built over the Passaic River around 1850 and replaced with an iron bridge during the 1890s. (Courtesy of the NSPCB archives.)

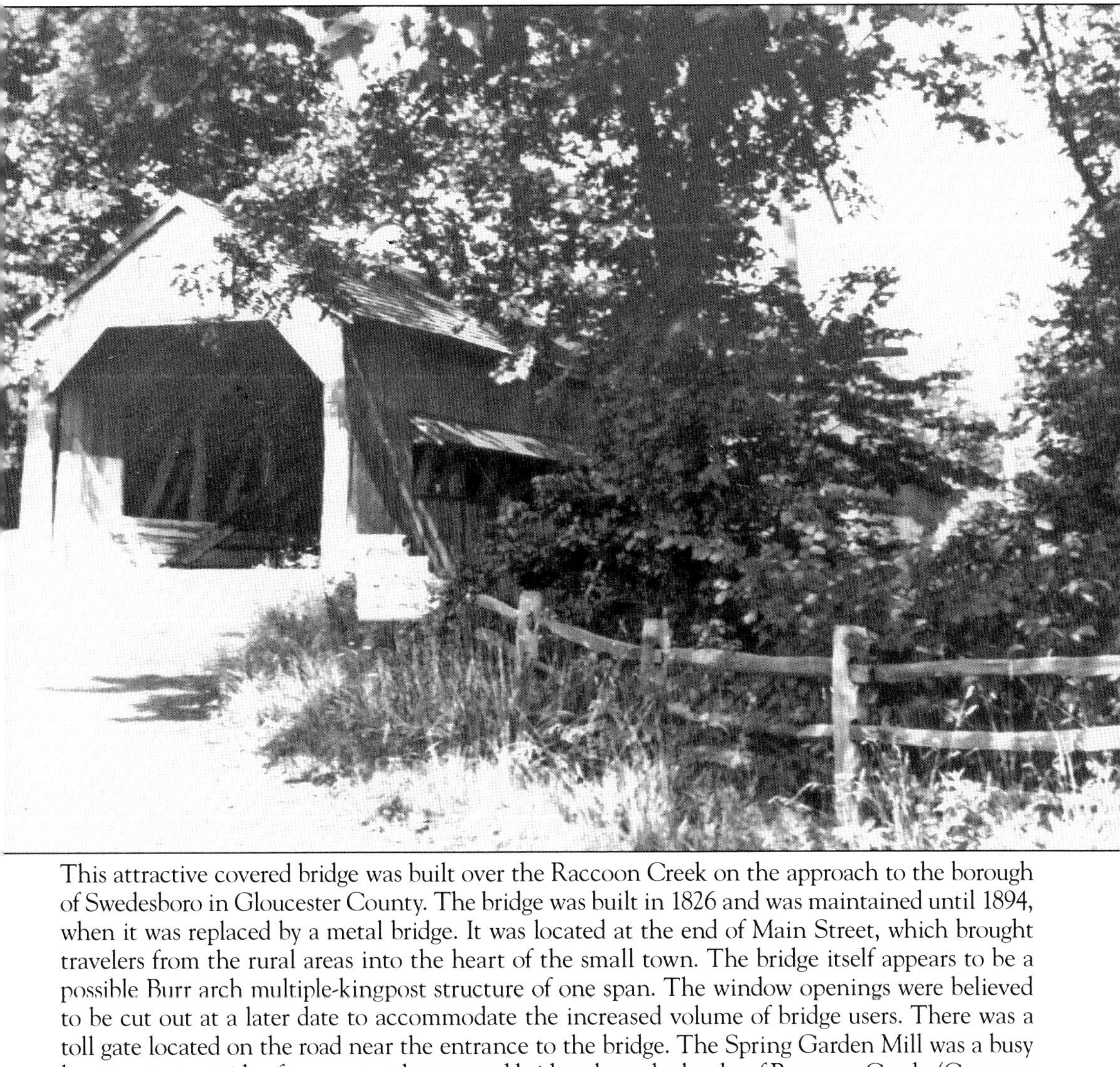

This attractive covered bridge was built over the Raccoon Creek on the approach to the borough of Swedesboro in Gloucester County. The bridge was built in 1826 and was maintained until 1894, when it was replaced by a metal bridge. It was located at the end of Main Street, which brought travelers from the rural areas into the heart of the small town. The bridge itself appears to be a possible Burr arch multiple-kingpost structure of one span. The window openings were believed to be cut out at a later date to accommodate the increased volume of bridge users. There was a toll gate located on the road near the entrance to the bridge. The Spring Garden Mill was a busy location just outside of town near the covered bridge along the banks of Raccoon Creek. (Courtesy of Gloucester County Historical Society.)

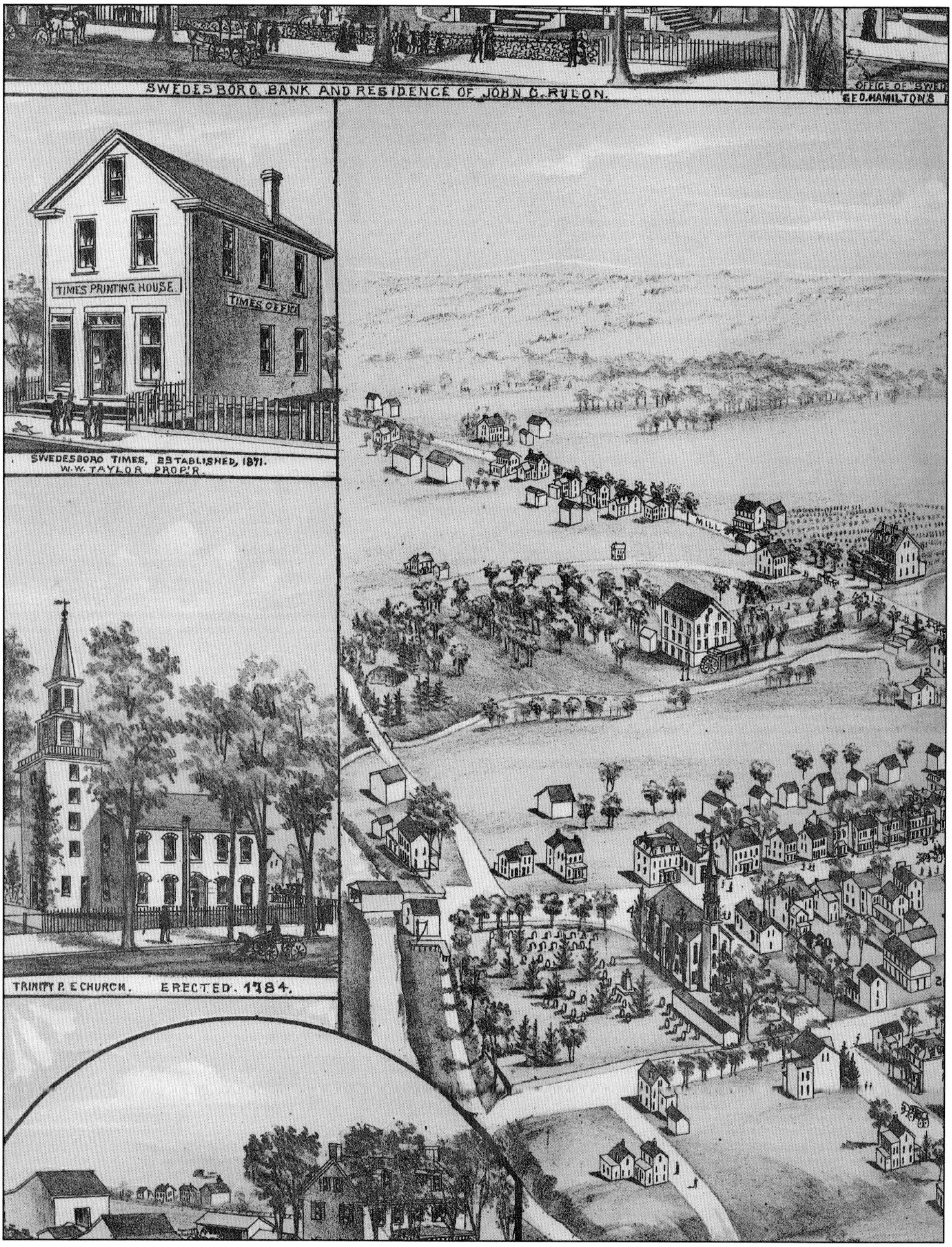

The work of T.M. Fowler, this 1886 bird's-eye-view map of Swedesboro is located at both the Swedesboro Public Library and the Gloucester County Historical Society. The image of the covered bridge is clearly seen on the far left side of the map, just outside the town itself. The road was then known as Main Street and some years later became King's Highway. In 1886, the covered bridge had no windows or openings. In an 1868 article of the then *Swedesboro Newspaper*, the editor told the story of kids playing by the covered bridge and then setting the roof on fire as a result of lighting off firecrackers. The old covered bridge had remained in service from its beginning in 1826 until 1894. Then the freeholders decided to replace it with a wide iron bridge and large walkways on both sides. (Courtesy of the Gloucester County Historical Society.)

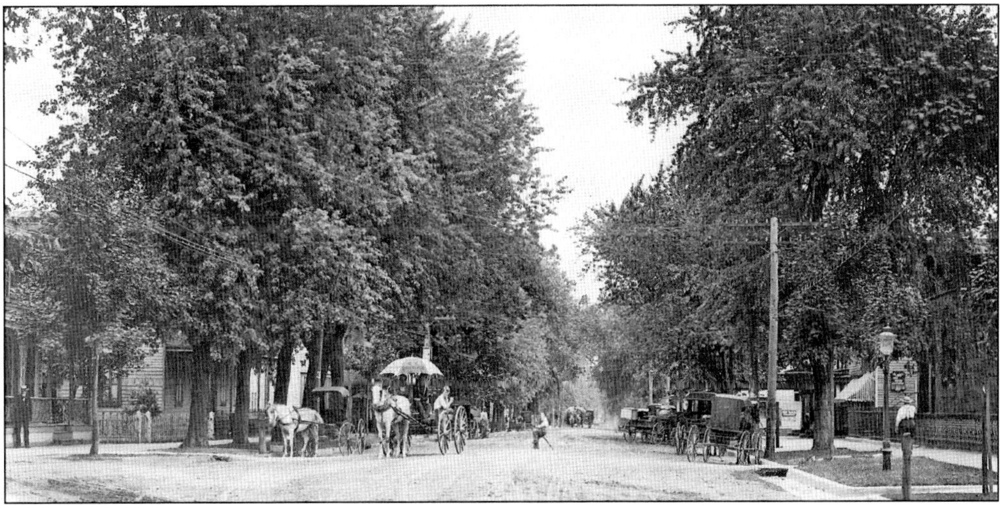

A group of Swedes and Finns settled the southern New Jersey town of Swedesboro in the mid-1600s. Mostly hunters and fishermen, the settlers began farming the area, rich with wood and fertile soil, and befriended the native Lenni Lenape as they expanded into New Jersey creek areas. Originally named Raccoon, the borough's name was changed to Swedesboro in 1765. (Courtesy of the Gloucester County Historical Society.)

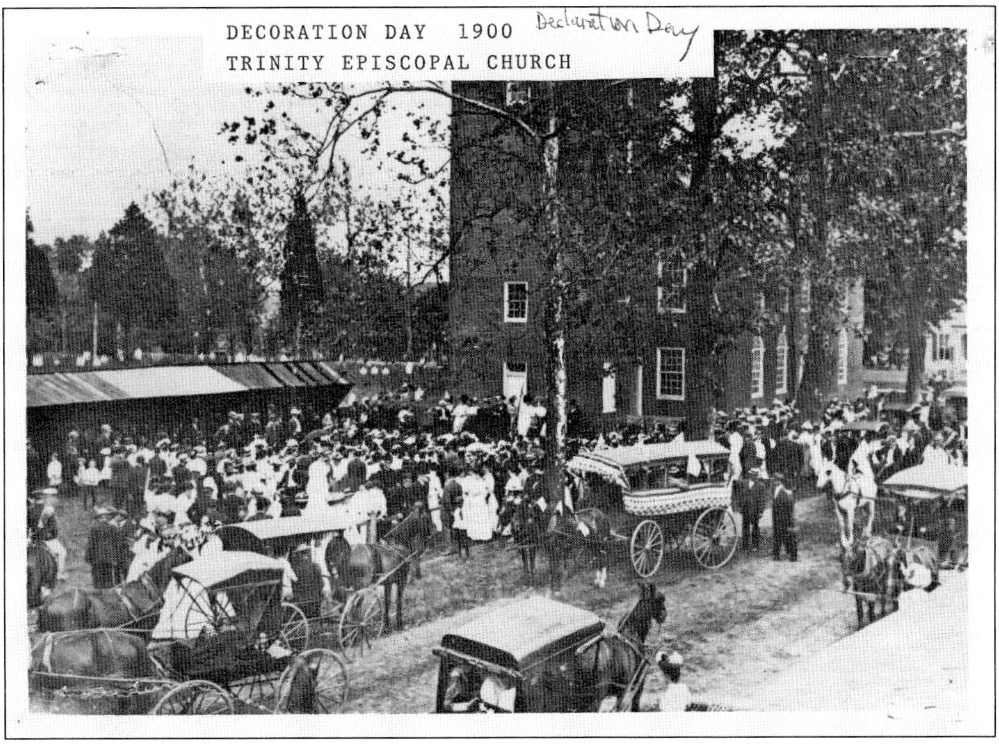

At Decoration (Declaration) Day festivities in 1900, the Gettysburg Address was read to the assembled crowd. Since its founding in 1638, Swedesboro has grown to a community of over 2,800. This historic borough has preserved its past in landmarks and eclectic architecture, including four local sites listed in the National and New Jersey State Registers of Historic Places. (Courtesy of the Swedesboro Public Library.)

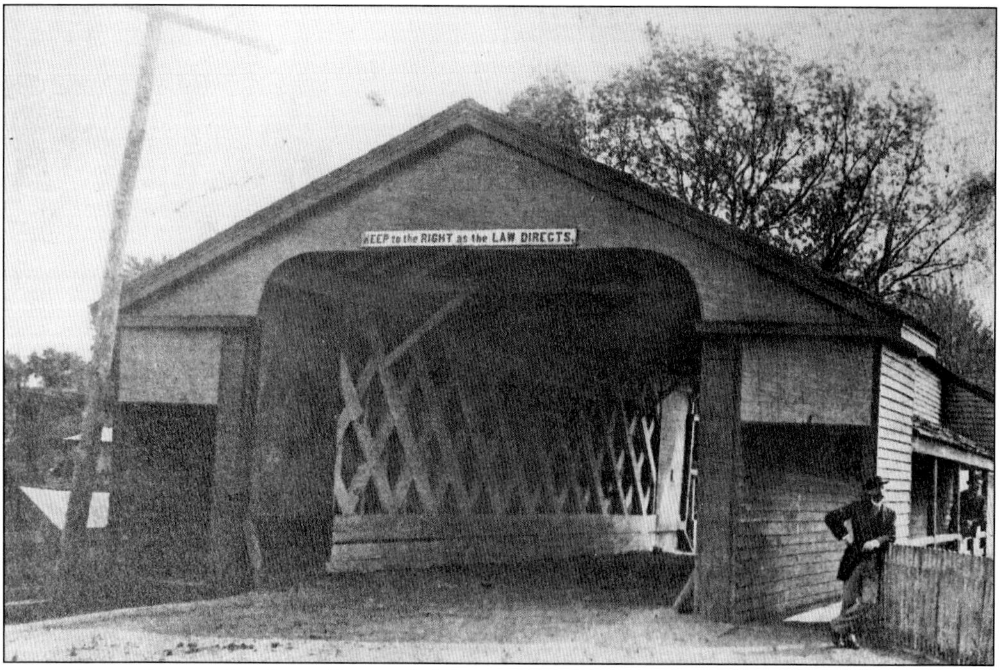

A gentleman lounges by the walkway of the covered bridge over Woodbury Creek in the city of Woodbury, Gloucester County. Built in the 1840s using the Town lattice truss, the bridge had a double-wide portal, providing for two-way traffic and covered walkways on both sides of the bridge. The sign reads, "Keep to the right as the law directs." (Courtesy of the Gloucester County Historical Society.)

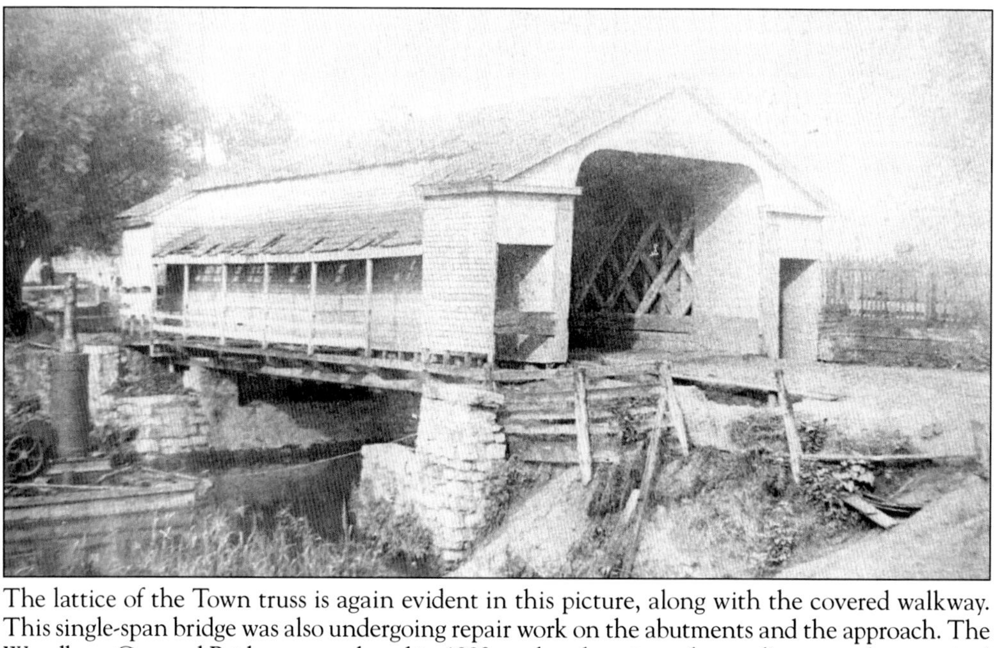

The lattice of the Town truss is again evident in this picture, along with the covered walkway. This single-span bridge was also undergoing repair work on the abutments and the approach. The Woodbury Covered Bridge was replaced in 1893, and at that time, the road's approach was raised about nine feet to allow for the passage of canal boats underneath. (Courtesy of the Gloucester County Historical Society.)

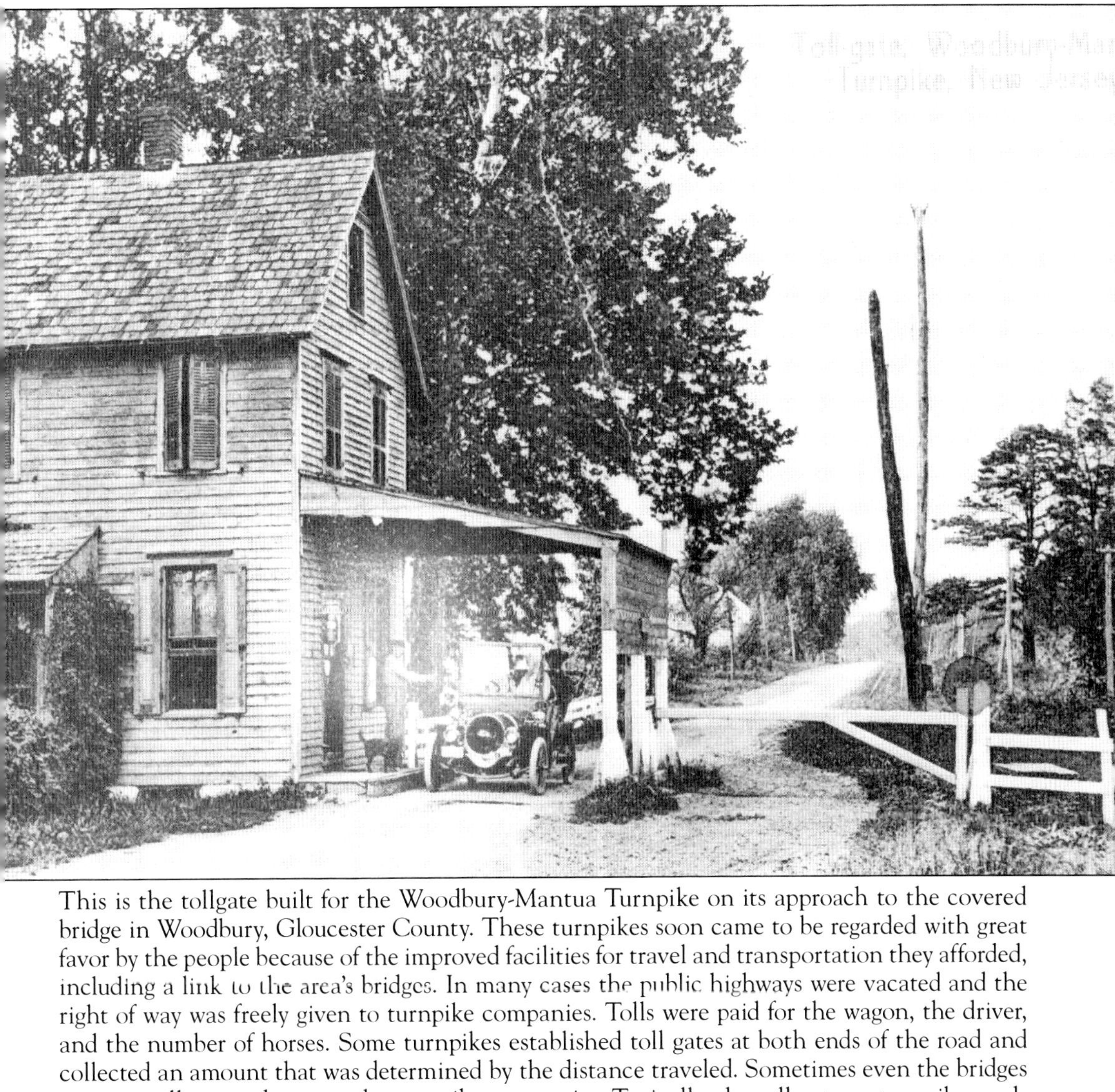

This is the tollgate built for the Woodbury-Mantua Turnpike on its approach to the covered bridge in Woodbury, Gloucester County. These turnpikes soon came to be regarded with great favor by the people because of the improved facilities for travel and transportation they afforded, including a link to the area's bridges. In many cases the public highways were vacated and the right of way was freely given to turnpike companies. Tolls were paid for the wagon, the driver, and the number of horses. Some turnpikes established toll gates at both ends of the road and collected an amount that was determined by the distance traveled. Sometimes even the bridges were actually turned over to the turnpike companies. Typically, the toll gate on turnpike roads was part of the toll house where the toll collector and his family members lived. (Courtesy of Gloucester County Historical Society.)

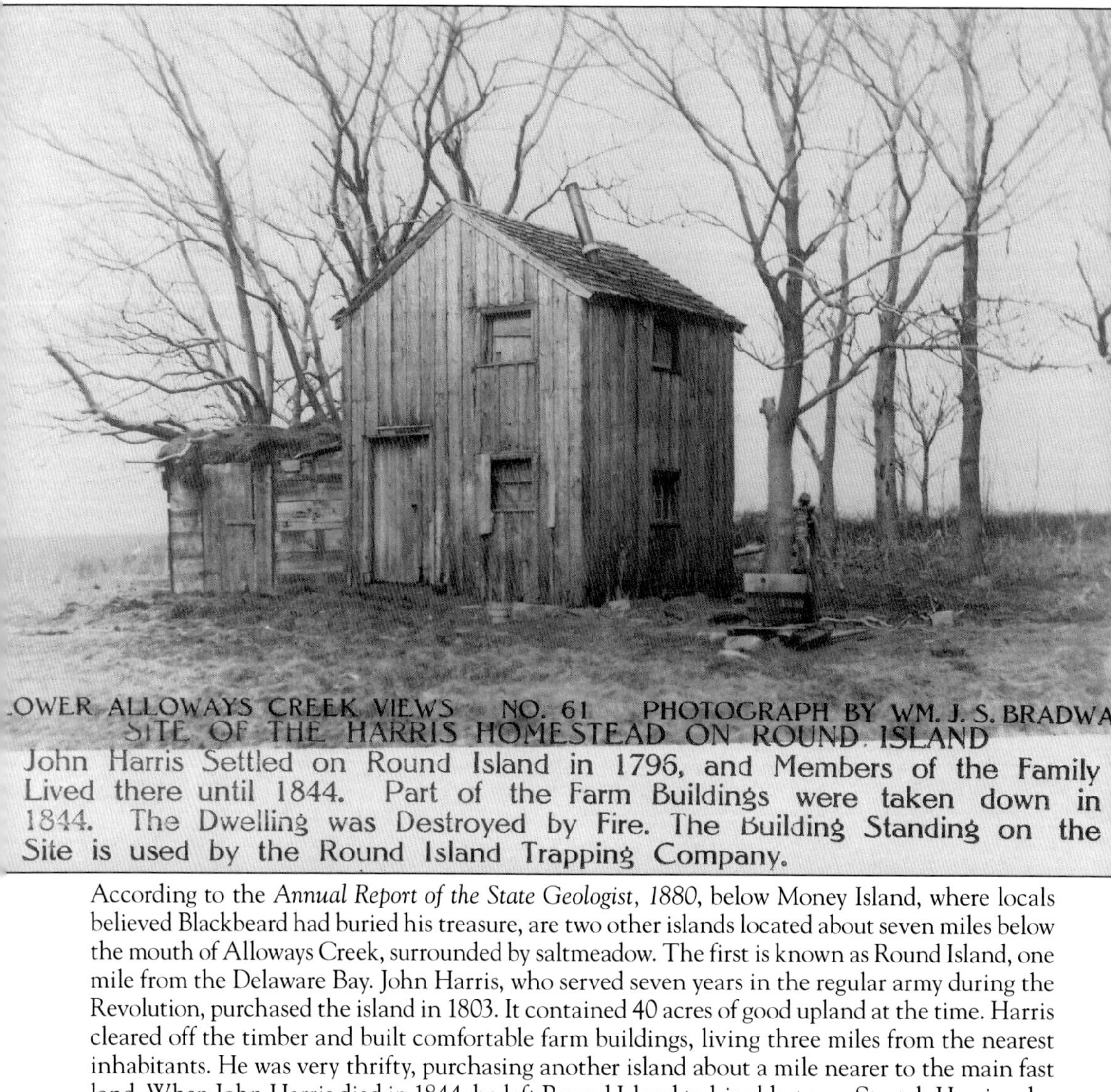

According to the *Annual Report of the State Geologist, 1880*, below Money Island, where locals believed Blackbeard had buried his treasure, are two other islands located about seven miles below the mouth of Alloways Creek, surrounded by saltmeadow. The first is known as Round Island, one mile from the Delaware Bay. John Harris, who served seven years in the regular army during the Revolution, purchased the island in 1803. It contained 40 acres of good upland at the time. Harris cleared off the timber and built comfortable farm buildings, living three miles from the nearest inhabitants. He was very thrifty, purchasing another island about a mile nearer to the main fast land. When John Harris died in 1844, he left Round Island to his eldest son, Stretch Harris, who was successful in accumulating so much wealth that he purchased a large farm on the mainland about 1.5 miles from the Hancock Bridge. (Courtesy of the New Jersey Historical Society.)

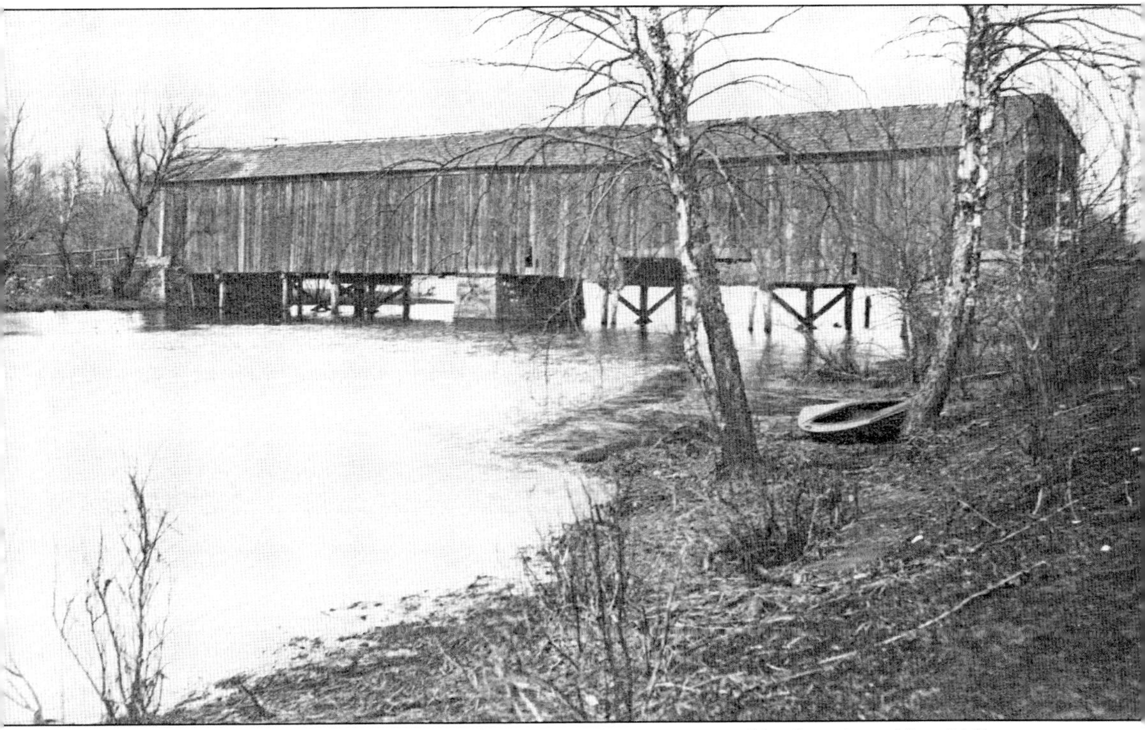

Three names are actually associated with this Mercer County covered bridge: Asay (Asay's) Spring, White Horse, and Abbottville. Asay Springs may not refer to an area but to actual springs that are located in the White Horse areas of present-day Hamilton, across the Crosswicks Creek from Bordentown. The reference to Abbottville relates to the proximity of the bridge location to the early Abbott's family plantation. Norman F. Brydon, in his 1970 book on New Jersey covered bridges titled *Of Time, Fire and the River*, wrote that the bridge was built in 1827, a single span, although there appears to be center support, about 80 feet long, a single driveway, no walkway, and solid sides without windows. The original drawbridge spanned the stream in June 1778 when the British soldiers attempted to cross the creek. In *Wasteland Wanderings*, Charles Conrad Abbott wrote in 1887, "But the present bridge, what is it? Like many another rude structure far away from the bustle of a town, it is the home of a host of creatures furred and feathered, for mice hide in its roof and muskrats in the abutment walls." (Courtesy of Todd Clark.)

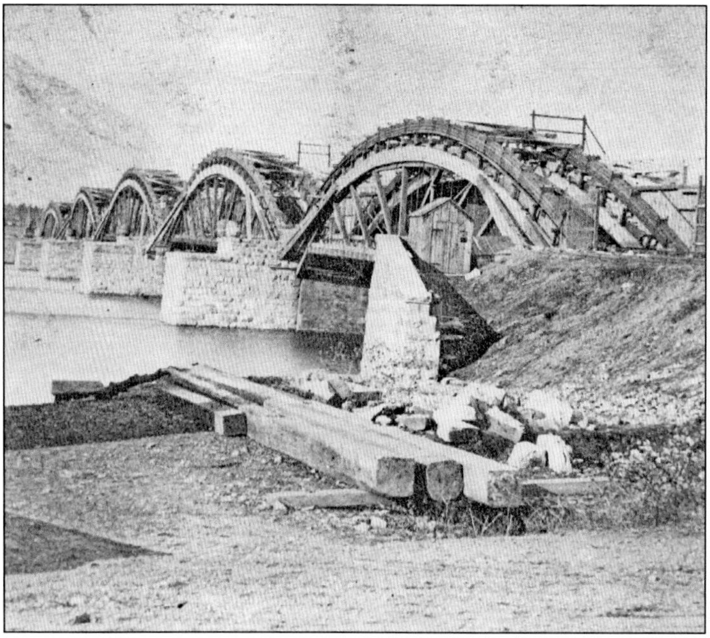

Chartered in 1798, the wooden bridge between Trenton, Mercer County, and Morrisville, Pennsylvania, was built by the Trenton Bridge Company and was the first across the Delaware River. The contract was made with Theodore Burr, the celebrated bridge architect and builder, and the bridge was completed in January 1806. (Courtesy of the Trentoniana Collection, Trenton Free Public Library.)

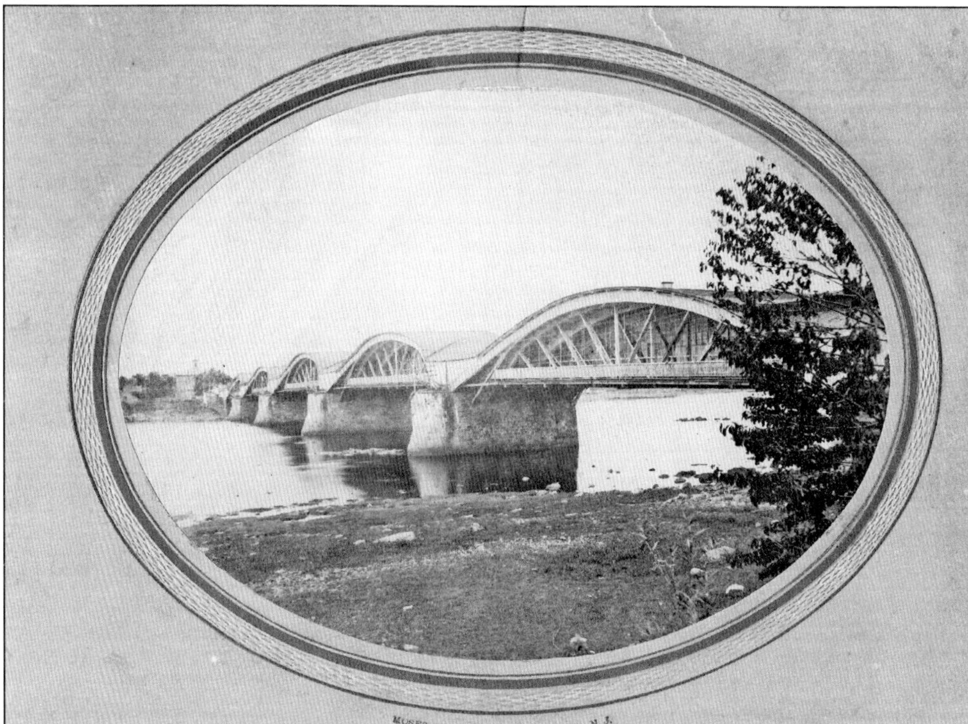

Its length was 1,100 feet, and its width was 36 feet through five spans. Before its completion, a freshet in the river proved the abutments were too low, so they were raised about one-fourth higher. The opening of the bridge on January 30, 1806, was made a festive occasion. The company president thanked Theodore Burr and others, and the celebration concluded with dinner, speeches, and toasts. (Courtesy of the Trentoniana Collection, Trenton Free Public Library.)

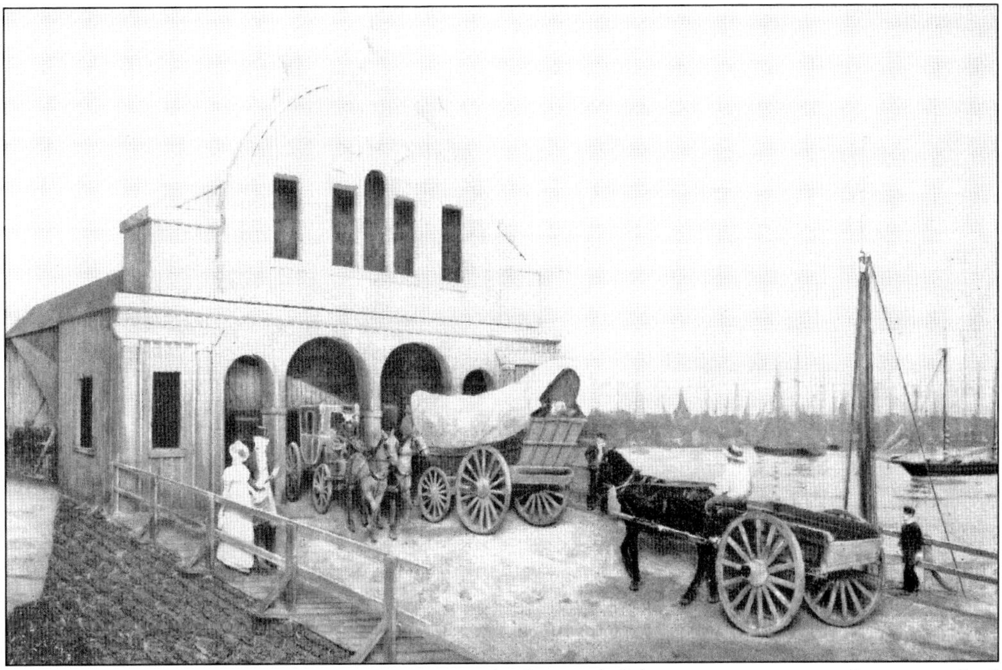

Originally a toll bridge, the Trenton Lower Bridge boasted elaborate entrances with great arched doorways over the carriage roads and footpaths. Balustrades of four feet in height protected pedestrians on the outside of the footpaths. Tolls were not cheap, with 75¢ charged for every coach or other pleasure carriage with four wheels and drawn by four horses. When drawn by two horses, the toll was 50¢. The road improvements enabled the expansion of travel from a single horse rider to stagecoaches for several people. Travelers referred to these coaches as "flying machines" because of their speed. A New Jersey first was a heavy farm wagon, or "Jersey Wagon," with huge wheels and a hooped cloth top that required four to six horses to pull it. (Both, courtesy of the Trentoniana Collection, Trenton Free Public Library.)

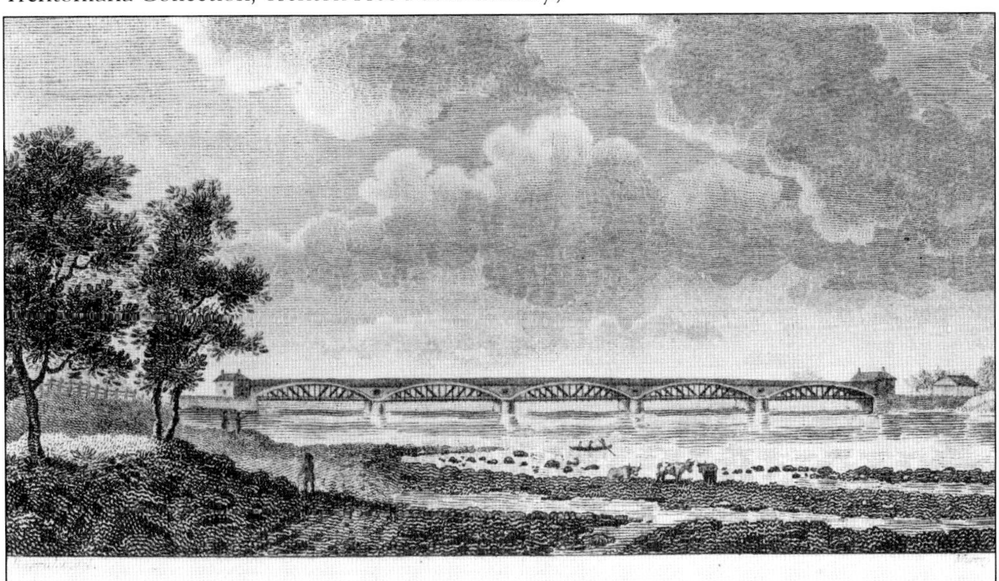

Trenton Bridge.

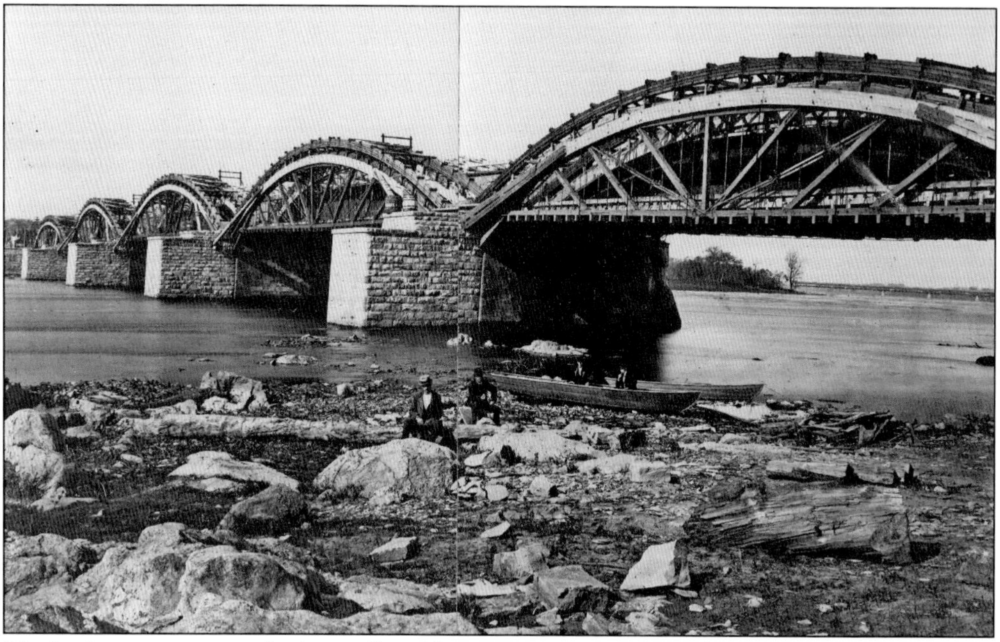

The bridge was a curious-looking affair. The superstructure of wood consisted of five arched rib beams. The floor was hung from iron chains, and to prevent it from swaying, wing arches and braces were used. Over the entire length of the bridge was a roof of cedar shingles, and the sides were also enclosed. The bridge rested upon the abutments and four piers, all of stone. The ends of the piers upriver were semicircular, gradually receding to the top, where they finished off in a half dome. These piers were 62 feet long and 20 feet deep. The bridge in South Trenton was also the first bridge in the United States to have been used in interstate railroad traffic. The bridge was replaced in 1876. (Both, courtesy of the Trentoniana Collection, Trenton Free Public Library.)

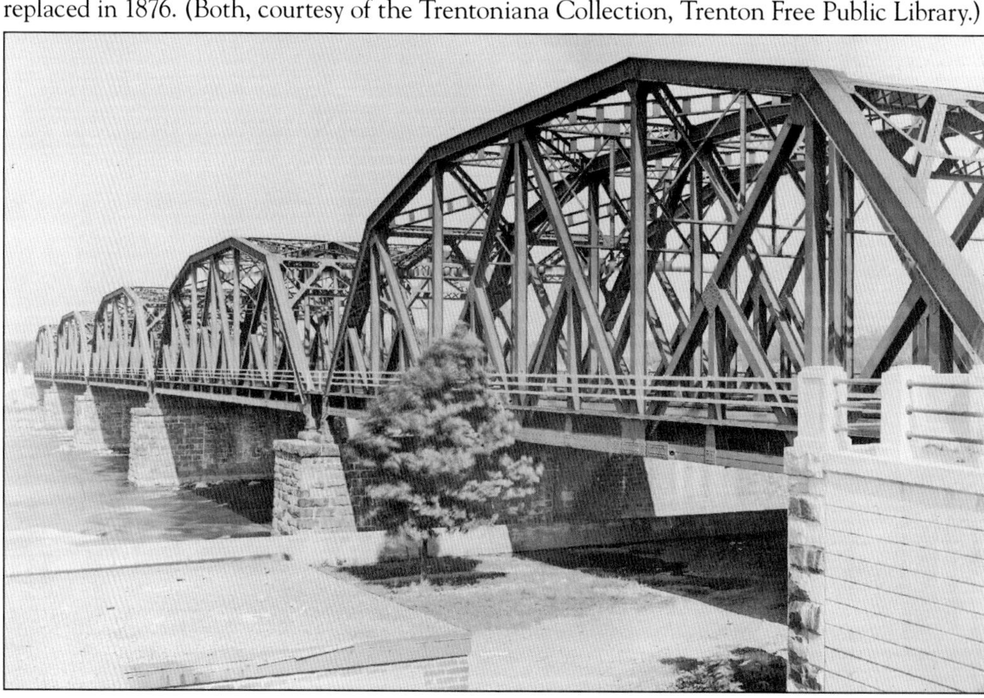

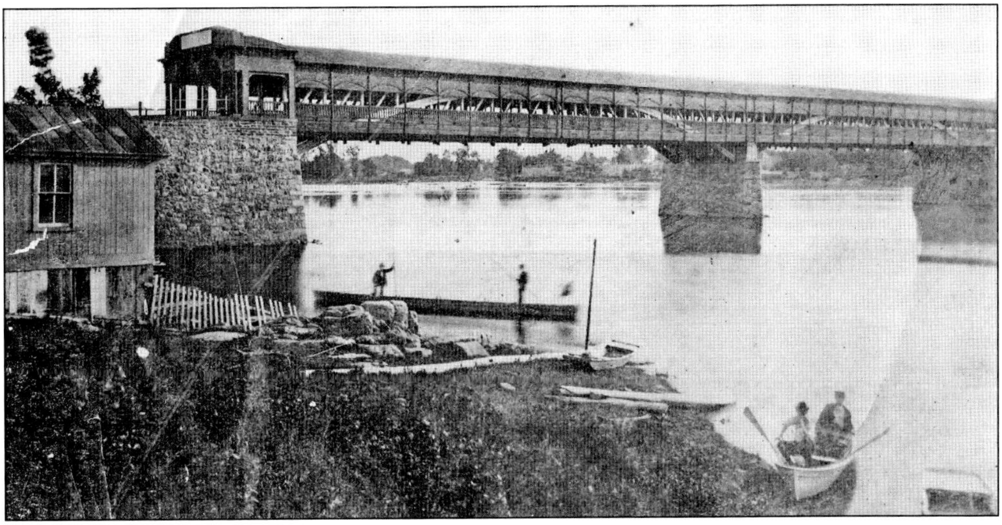

The long-anticipated second bridge built across the Delaware River at Trenton, known as the City Bridge or Calhoun Street Bridge, was opened to traffic on July 1, 1861. The need for this bridge was felt among the residents of Trenton and Upper Bucks County in Pennsylvania for more than two decades before it was built. (Courtesy of the Trentoniana Collection, Trenton Free Public Library.)

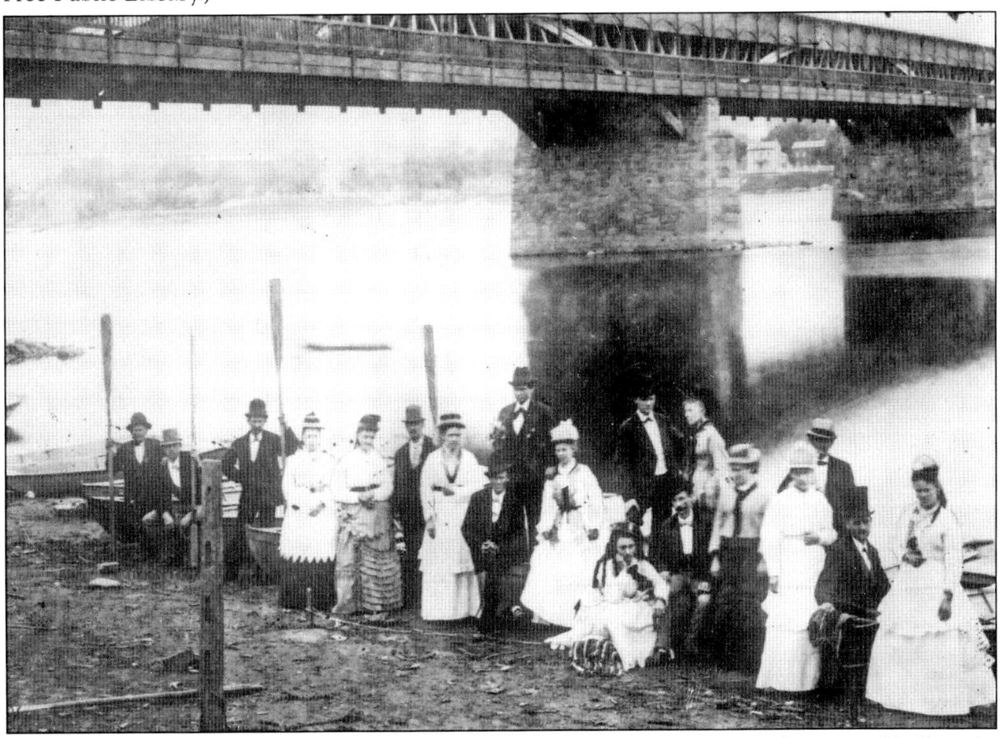

This photograph from the *Sunday Times-Advertiser* captures the "Trenton beaux and belles of a long vanished day." A Sunday rowboat ride on the river, drifting leisurely under the covered bridge, would culminate in a picnic on the shoreline of the Delaware. The Calhoun Street Bridge was hoped to attract the trade of a portion of Bucks County to Trenton. (Courtesy of the Trentoniana Collection, Trenton Free Public Library.)

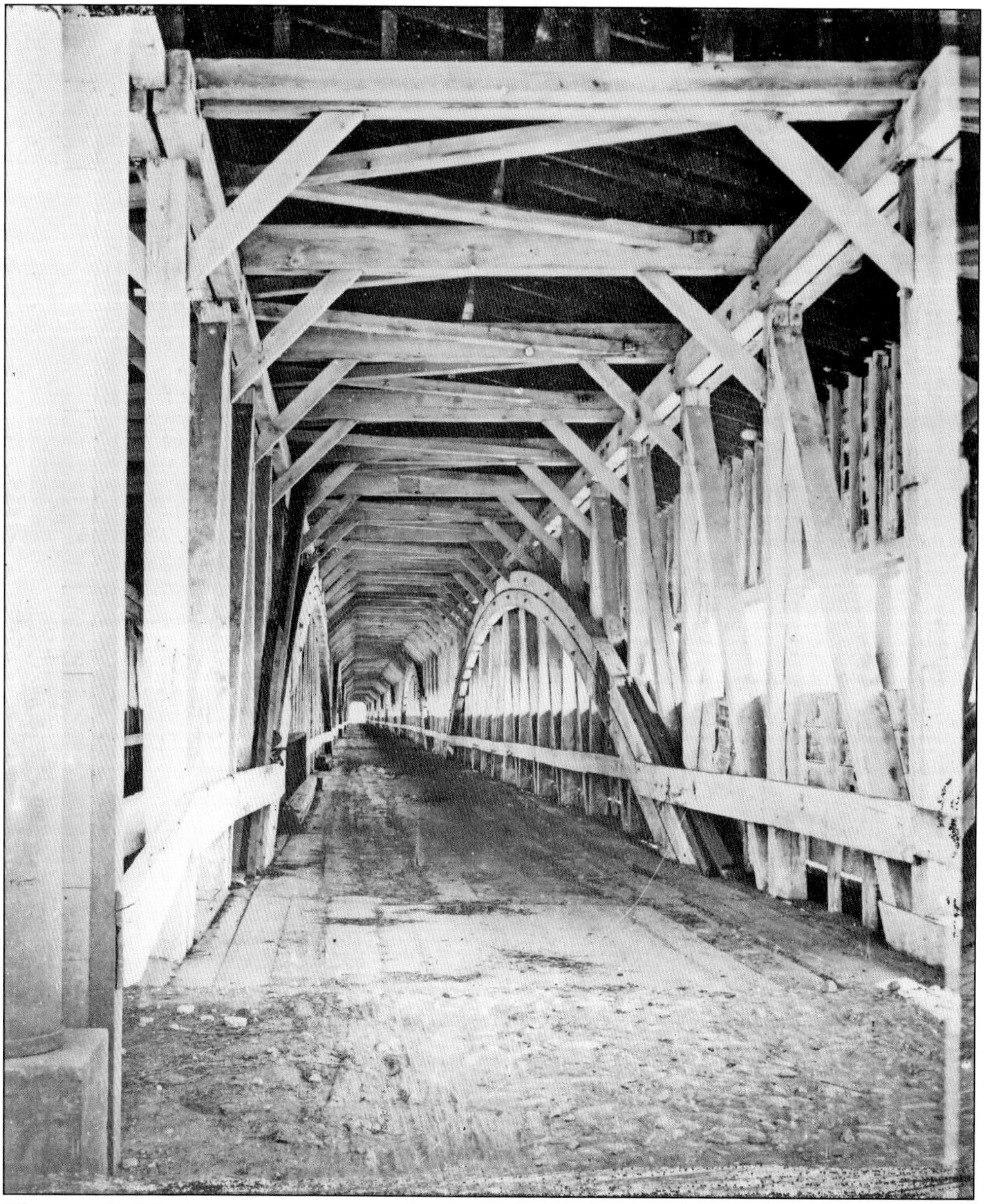

This wonderful picture presents one side, or one lane, of the Calhoun Street Bridge over the Delaware River in Trenton. A two-lane covered bridge is called a double-barrel bridge. Most covered bridges were only wide enough for a single lane. The bridge had a length of 1,274 feet, with seven spans built in a Burr-arch and multiple-kingpost truss on six piers in the river and the two abutments. Along with the two lanes, there were also two walkways for pedestrians. The Trenton City Bridge Company was incorporated by Pennsylvania in 1840 and New Jersey in 1842 to select a site and build the bridge. The company, however, did not get organized until about 1846. Finally, in 1859, a site was determined, and by early October of that year, work began. It was built higher above the normal water level of the river than any of the river bridges. Opening day was held on July 1, 1861. Over 200 foot passengers and some 100 vehicles crossed over the bridge that day. (Courtesy of the Trentoniana Collection, Trenton Free Public Library.)

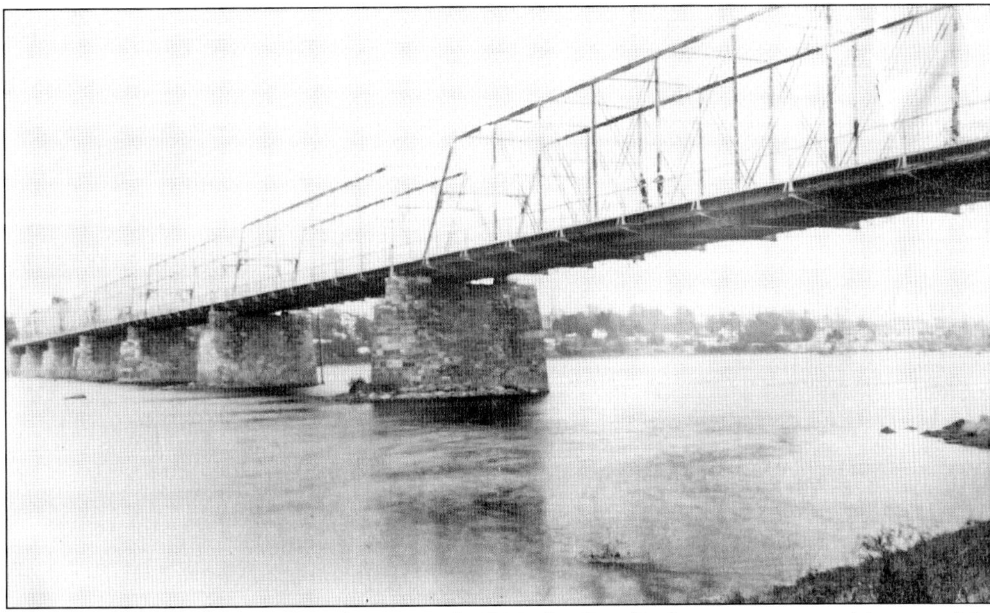

Early in the evening of Wednesday, June 25, 1884, the City Bridge was consumed in one of the most spectacular fires the residents of the Delaware Valley had ever witnessed. A carelessly dropped cigar or cigarette butt is believed to have been the cause. Soon after the destruction, the Trenton City Bridge Company planned to replace the bridge with an open iron structure. (Courtesy of the Trentoniana Collection, Trenton Free Public Library.)

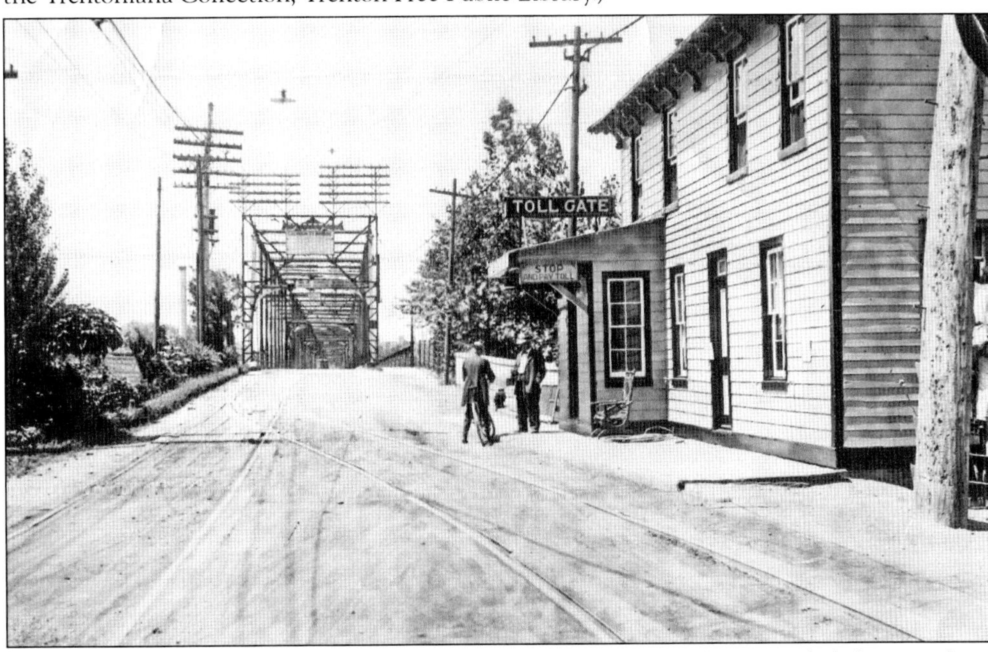

The men who organized the companies to build the covered bridges regarded the completed spans exactly as most financiers and developers regard all other structures today—as a means of achieving wealth. State charters provided that the tolls charged should be sufficient to pay a 6-percent return. Here, a man on his bicycle pays his toll at the City Bridge. (Courtesy of the Trentoniana Collection, Trenton Free Public Library.)

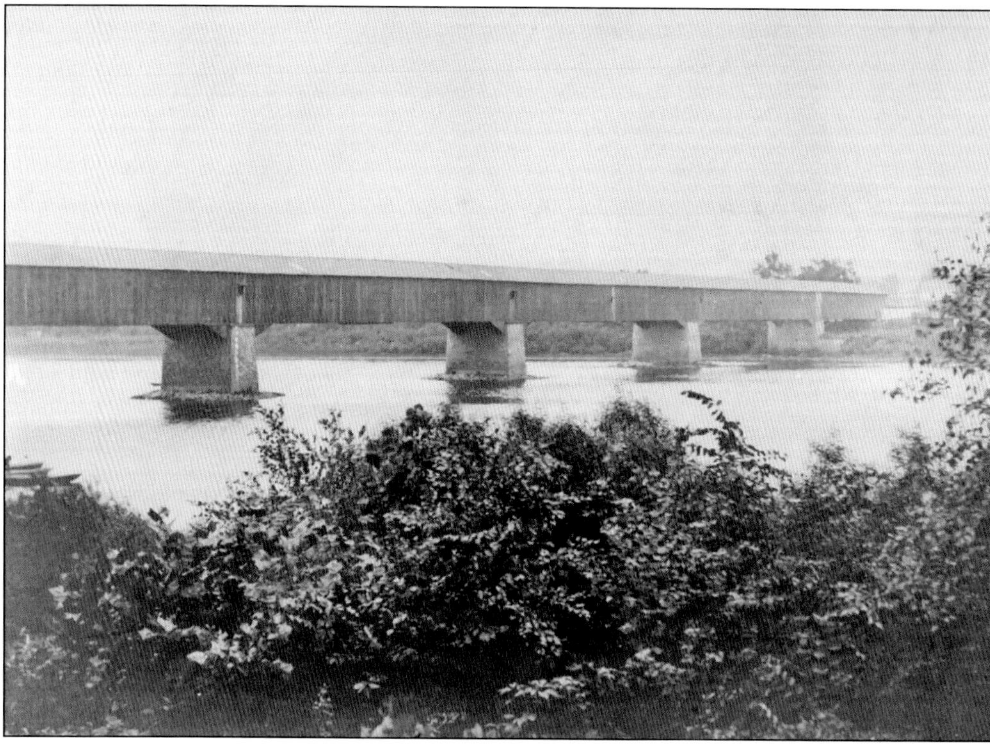

In Mercer County, above, the Yardley-Delaware Bridge Company built a new wooden covered toll bridge in 1835 over the Delaware River. Stone abutments and five stone piers supported a six-span, Burr-arch structure measuring over 900 feet in length. During a major flood in early 1841, the bridge lost three of its six spans but was rebuilt in the same style. Typical of the devastation caused by severe flooding on the Delaware, the image below highlights the result of the rampaging water and the objects carried with it. The flood would not only wash out the bridges but lift them off their abutments and piers to crash into the next bridge downriver. (Above, courtesy of Trentoniana Collection, Trenton Free Public Library; below, courtesy of Todd Clark.)

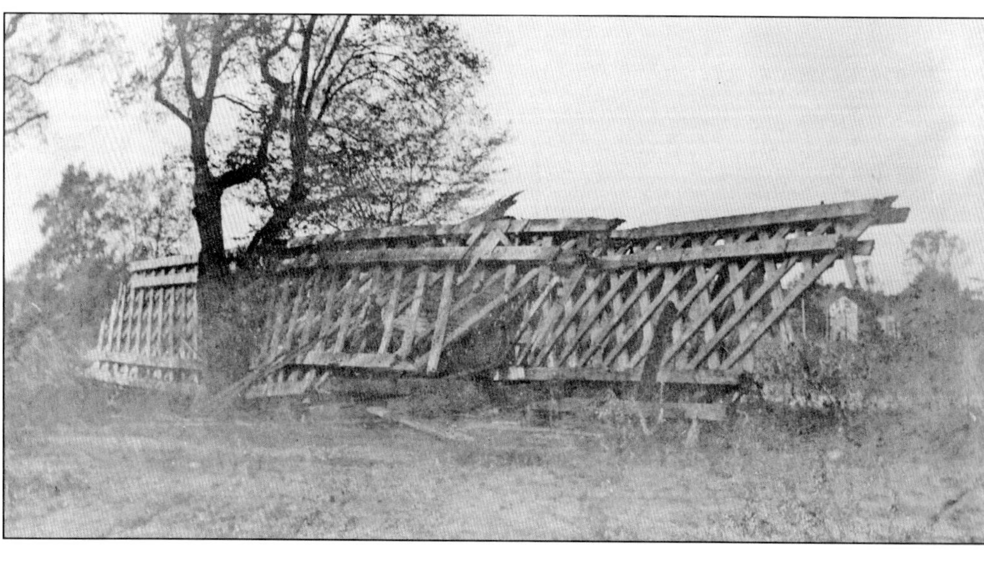

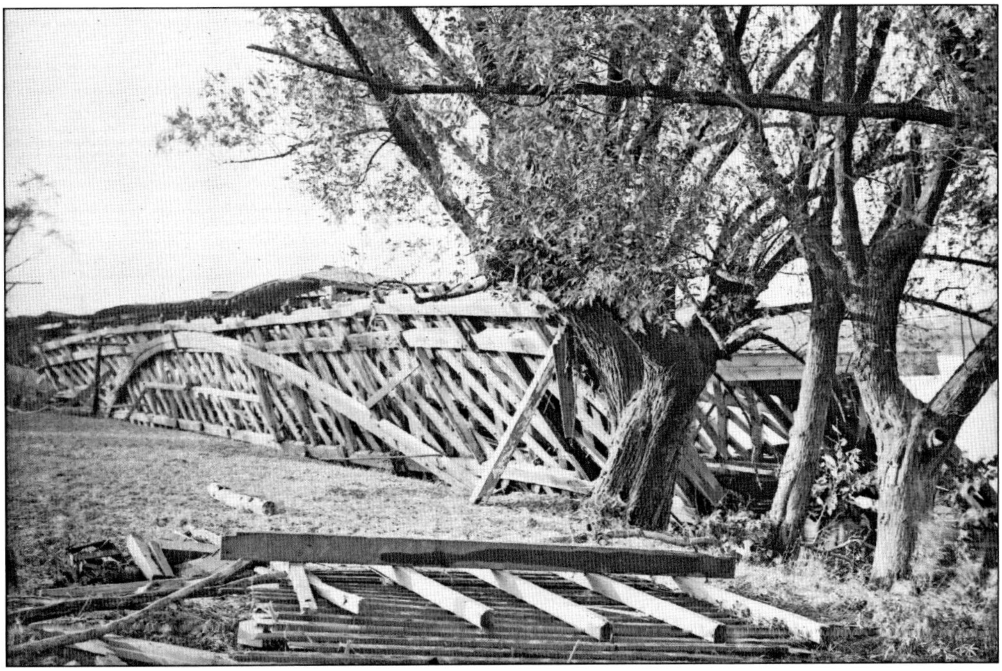

The Yardley-Wilburtha Bridge enjoyed six decades of use between two prosperous towns. Then, in 1903, a disastrous flood hit the Delaware River and swept away and destroyed its bridges. This bridge was not spared from this disaster and was torn apart. The bridge company now chose a new plan and had a six-span, Warren-truss steel bridge built on the original stone foundations. In 1922, the Joint Bridge Commission purchased the bridge, and tolls were eliminated. Record floodwaters in 1955 besieged the river and caused the destruction of three of the six spans of the bridge. Because of plans for a new bridge upriver, the Yardley-Wilburtha Bridge was never rebuilt. (Above, courtesy of Todd Clark; below, courtesy of the Trentoniana Collection, Trenton Free Public Library.)

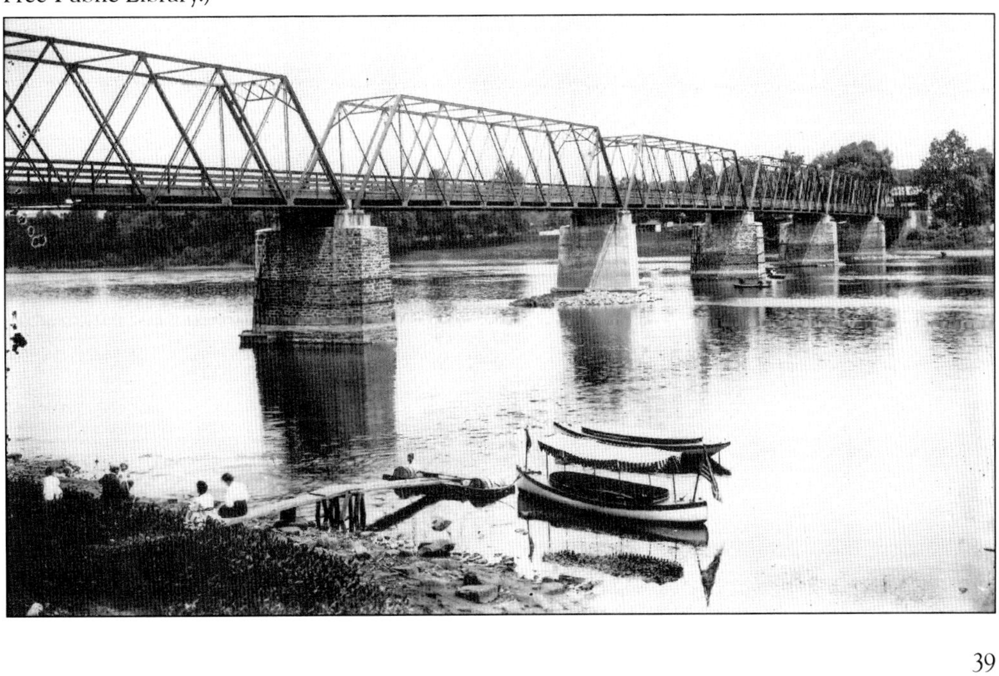

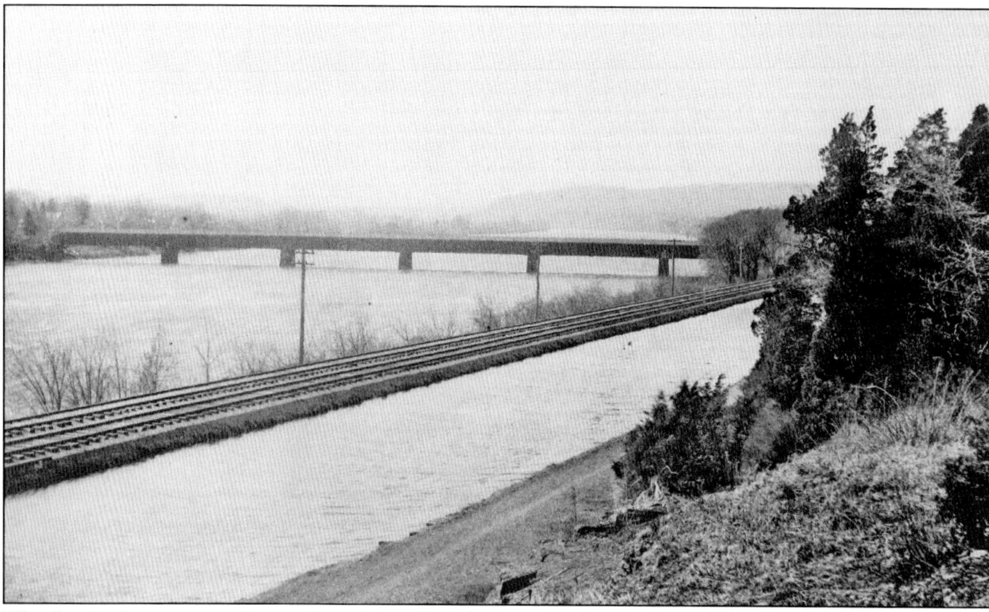

This bridge location is often referred to as Washington's Crossing in Mercer County. On Christmas night in 1776, George Washington led his army across the Delaware River here to surprise the enemy troops barracked at Trenton. The ferry docks belonged to a Samuel McKonkey, but large Durham boats actually carried everyone across the river. (Courtesy of the Trentoniana Collection, Trenton Free Public Library.)

RATES OF TOLL.

Vehicle, Drawn by Beasts of Burden, per head,	10 cents
Single Horse, or Mule, and Rider,	10 "
Led, or Driven Horse, or Mule,	5 "
Cattle, per head,	3 "
Sheep or Swine, per head,	2 "
Foot Passenger,	2 "
Light Automobile,	10 "
Heavy "	20 "
Bicycles, per head,	2 "
Wheelbarrow,	4 "
Steam Thresher, per horse,	10 "
Steam Engine, " "	10 "
Steam Roller,	$1.00
Traction Engine,	1.00

By order of

WASHINGTON'S CROSSING DELAWARE BRIDGE COMPANY.

April 1st, 1905.

The covered bridge was first chartered in 1831. Both New Jersey and Pennsylvania sought financial security of the company before the bridge was built. Opened in 1834, the bridge had six spans with a lattice truss, a length of 875 feet, and a width of 17 feet. Tolls kept the bridge a safe investment, but the great flood of 1903 ultimately destroyed it.

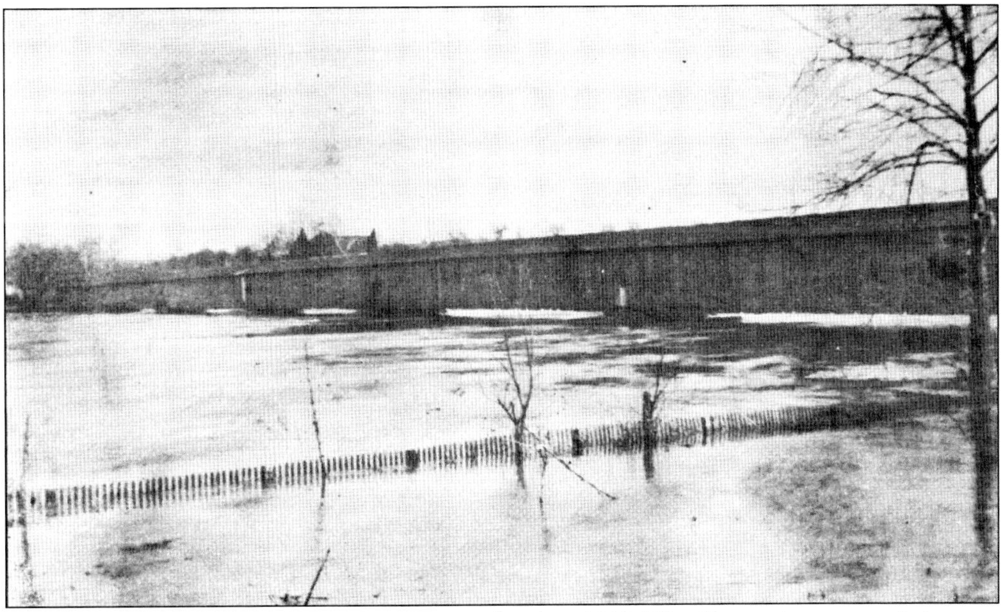

Floodwaters rose at Washington Crossing in October 1903. The *Trentonian* reported that, up and down the East Coast, there was rain and plenty of it. Along the river valley was a disaster. Helpless, residents of Trenton, Lambertville, and other towns watched as the usually calm waters overspilled their banks, reaching levels from 10 to 28 feet. Nine wooden bridges linking the two states were lifted off their stone piers and smashed to splinters. Ten days after the flood, a horrible sequel occurred when two trains collided on their way to repair the bridge at Washington Crossing. (Above, author's collection; below, courtesy of the Trentoniana Collection, Trenton Free Public Library.)

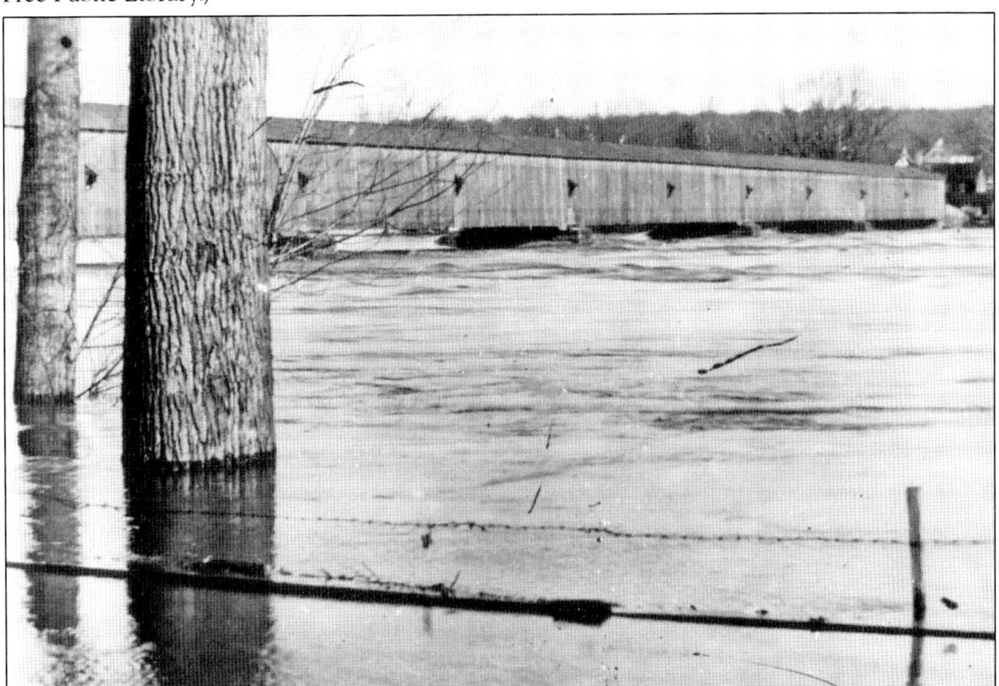

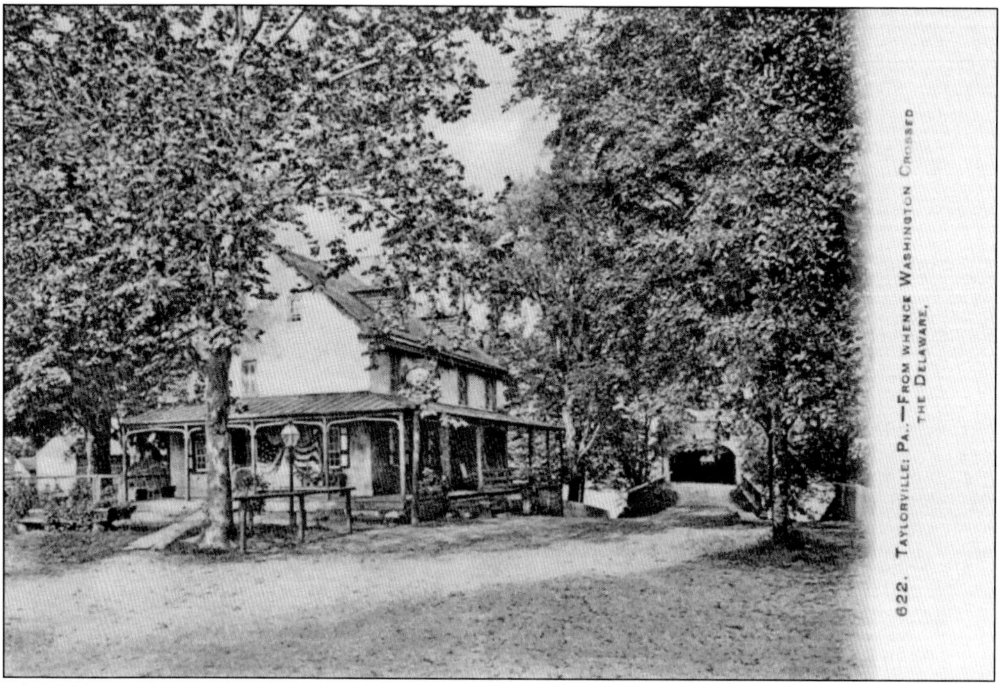

The following is an excerpt from the *Pictorial Field Book of the Revolution* by Benson J. Lossing: "The feeder for the Delaware and Hudson Canal and the artificial channel made along the river bank for water power at Trenton, are crossed and paralleled by the road all the way to Yardleyville. Taylorsville is on the Pennsylvania side at M'Conkey's Ferry. A noble bridge six hundred feet long here spans the river. It is supported by eight piers, eighteen feet above the water when the stream has its usual depth. The bridge is of timber, the piers of solid masonry, with an icebreaker on the upper side. The view here given is from below the bridge on the Pennsylvania side, looking northeast and exhibits the Jersey shore." (Above, courtesy of Todd Clark; below, author's collection.)

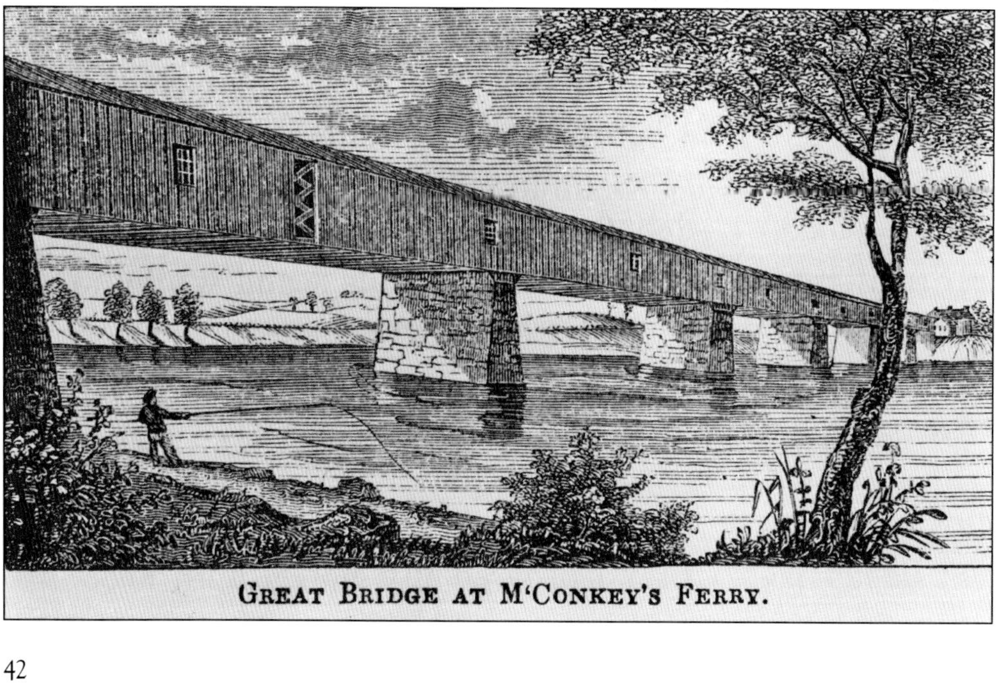

Great Bridge at M'Conkey's Ferry.

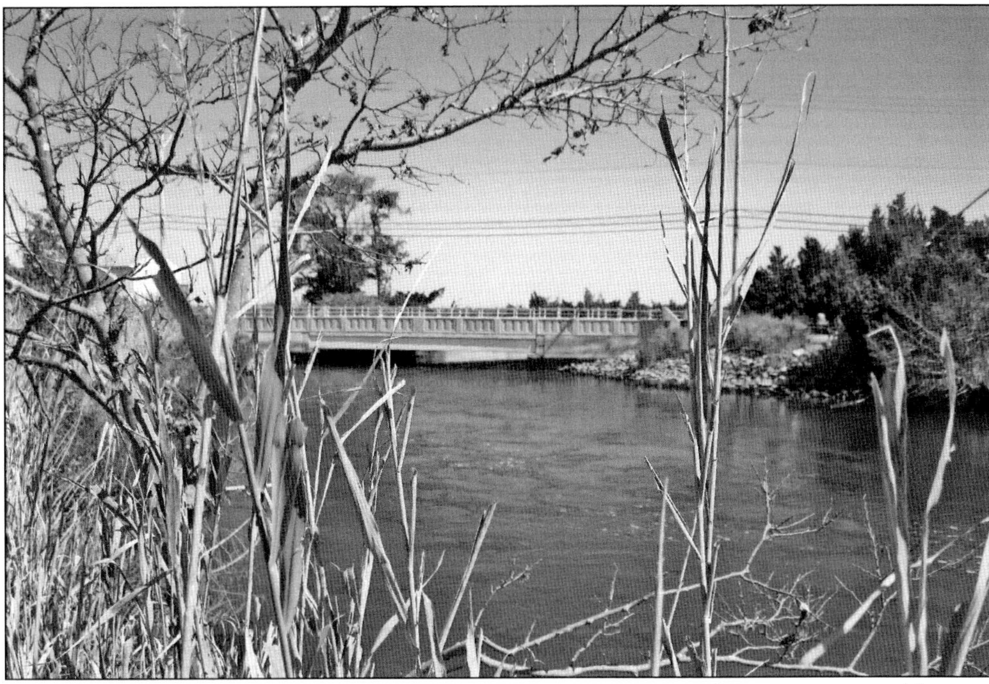

As written in the *Cultural Resources Digest*, a New Jersey Department of Transportation publication, a causeway was built across Dennis Creek in Cape May County, which brought profound changes to the area of Dennisville. Between 1844 and 1885, a covered bridge provided an important link to nearby villages. The most recent bridge crosses on that exact location (above). Standing on that bridge and looking west, a barren open marshland (below) betrays the history of Dennis Creek Landing. By the 1870s, this was one of the more important ports and shipbuilding centers on the Delaware Bay, producing substantial numbers of schooners and other large sailing vessels. In 1890, a 707-ton ship—at 150 feet long, 34 feet wide, and drawing 17 feet of water—negotiated a passage down this creek to the Delaware Bay. (Both, courtesy of Jack Connolly.)

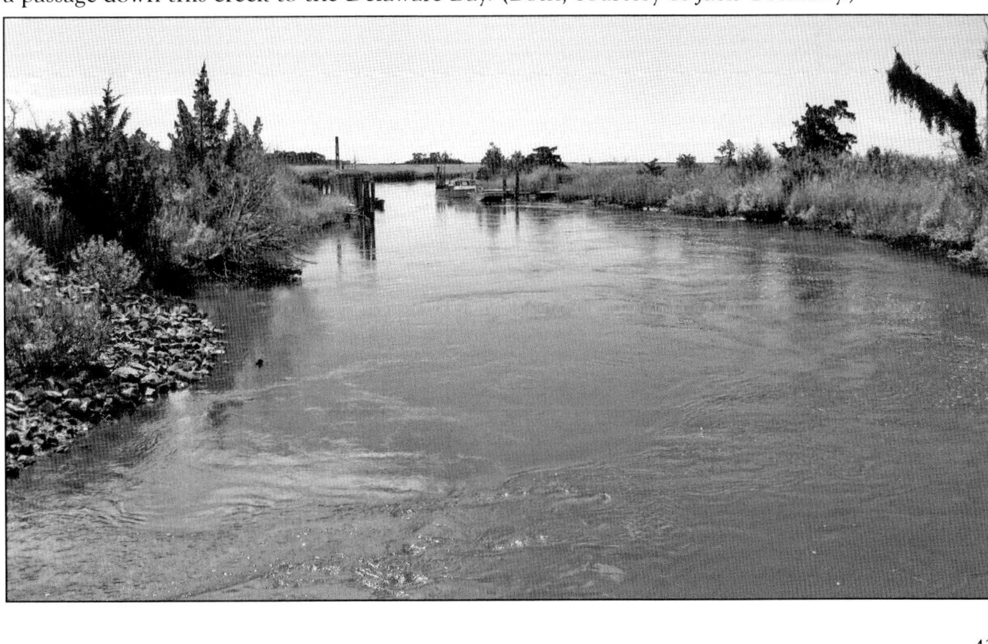

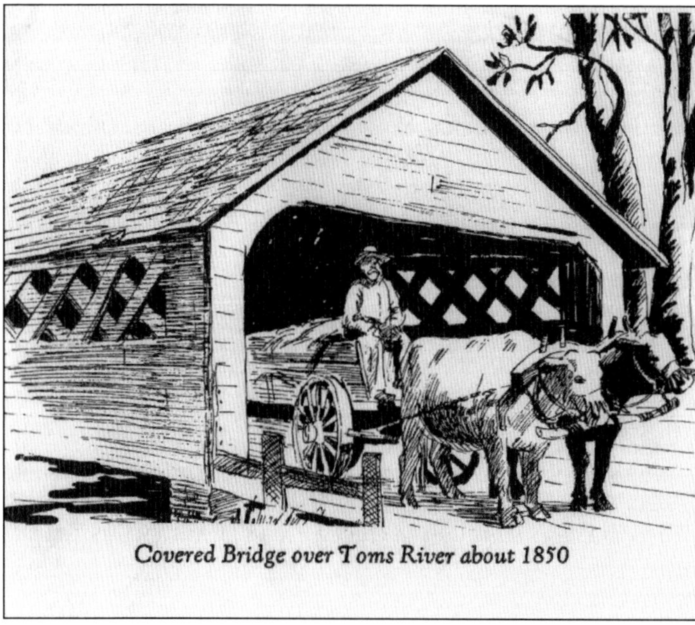

This illustration by Wini Smart is found in a 1967 work by Pauline Miller. The Toms River Covered Bridge, Ocean County, connected with South Toms River. Work on the new Town lattice bridge began in October 1850. The schooner *Moses Johnston* delivered the lumber for the bridge. This covered bridge served the community until 1872. (Courtesy of Wini Smart and Ocean County Historical Society.)

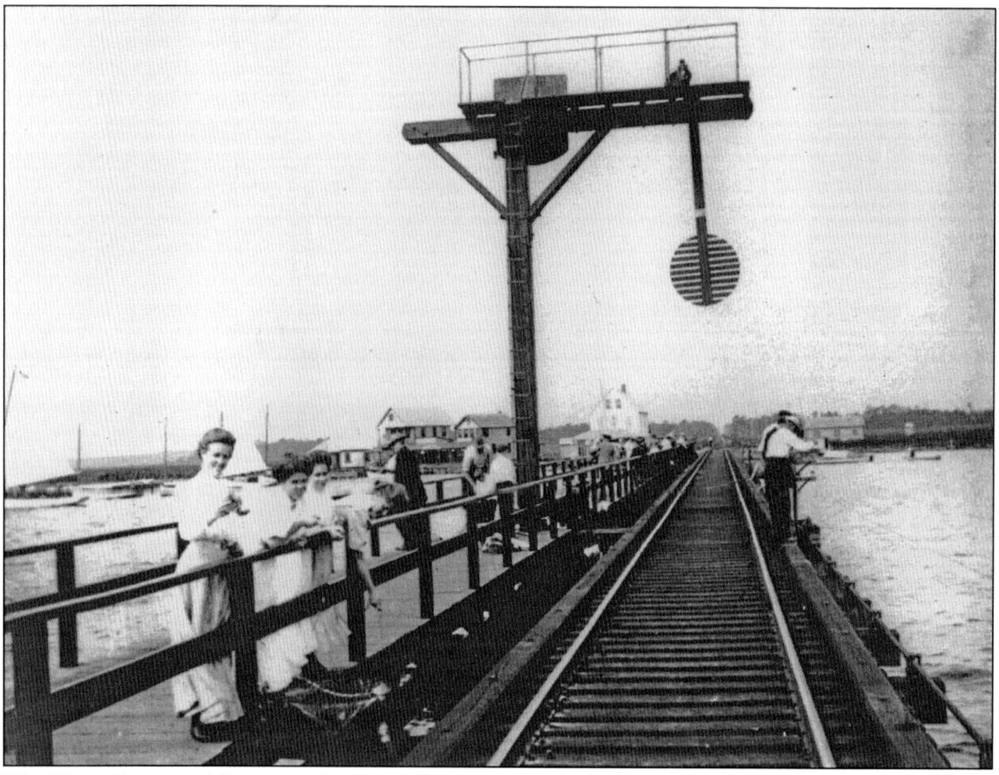

The Toms River and Barnegat Railroad Company, according to information on file from Wilson Brothers and Company, built a single track railroad bridge across Barnegat Bay between Good Luck Point and Seaside Park, Ocean County. The entire structure was 7,050 feet long and included a 164 foot Howe-truss swing bridge allowing boat passage. (Courtesy of Ocean County Historical Society.)

Three

The Southern Shore and Shore Regions

An extremely popular destination for many thousands of visitors and vacationers, these two regions include Atlantic, Cape May, and Cumberland Counties as the Southern Shore and Monmouth and Ocean Counties as the Shore Region, or simply, the Jersey Shore. Given the topography of much of the region, the necessity for many bridges would seem apparent. But practicality called for open wooden-truss bridges with an occasional Howe-truss drawbridge. The old Atlantic City Railroad drawbridge, dating to the mid-1850s, was such a structure that brought passengers to a train shed on Atlantic Avenue.

The Toms River Covered Bridge stands as one of the best examples of solid and determined historical research. The Ocean County Historical Society (OCHS) took action on hearsay stories and turned up authenticating archival proof for the true existence of this bridge over the Toms River. Its chronology begins with the arrival of lumber for the bridge on the schooner *Moses Johnston* at the port of Toms River in 1850. Already, in 1853, complaints were being made about "the practice of boys collecting on the covered bridge and crowding it to such an extent as to render passing and re-passing almost impossible," according to the *Ocean Emblem* of April 7, 1853. On July 4, 1872, the *New Jersey Courier* reported that "the ancient covered bridge" across Toms River was rapidly disappearing in preparation for a new bridge, while a pontoon bridge was temporarily in place "to facilitate the public travel." Finally, on August 8, 1872, again the *New Jersey Courier* reported, "It will not be many days before we can travel across on a firm reliable iron bridge."

With New Jersey's covered bridges simply and sadly nonexistent today, it is the hard and steadfast work of dedicated people, most of them volunteers, such as those at Ocean County Historical Society, to discover and authenticate these old treasures.

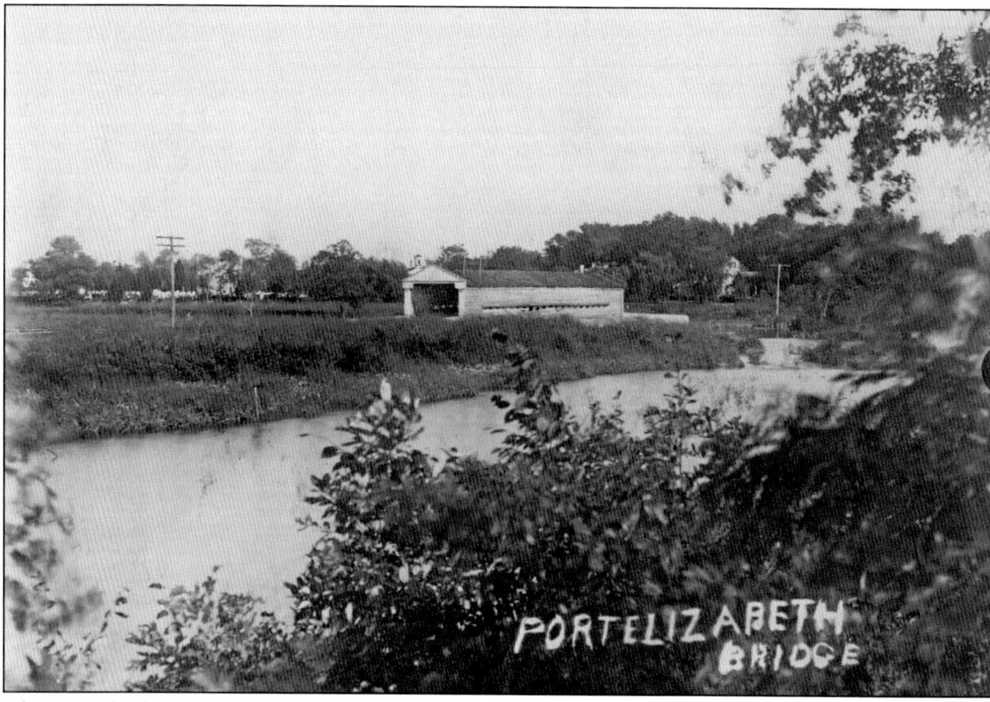

The Port Elizabeth Covered Bridge crossed one of the Manumuskin Creek channels in Cumberland County. According to the *History of Port Elizabeth* by F.W. Bowen, in 1821 a bridge was built over Manumuskin Creek in Port Elizabeth, at eight rods long, 18 inches high above all tides in the creek, and 20 feet wide in the clear. In 1838, a new covered bridge was built on the truss plan, at 60 feet long and 20 feet wide. Amos Campbell, a contractor, built it for a sum of $2,350. Port Elizabeth was first known as Maurice River Town, and Elizabeth Clark Bodley, who owned the land, laid out the streets in 1771. (Both, courtesy of Todd Clark.)

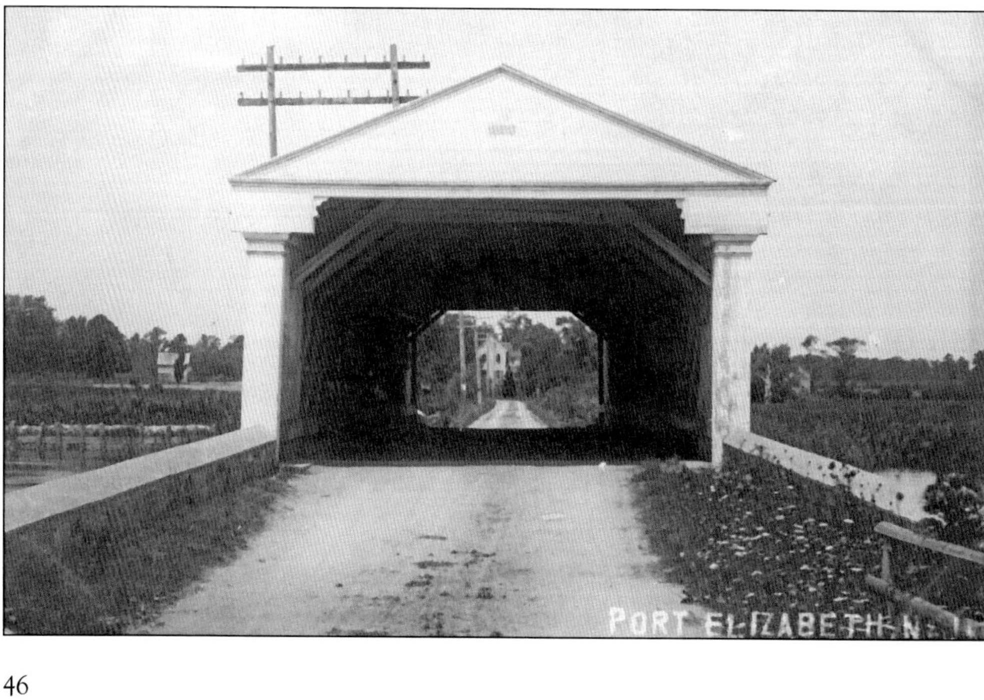

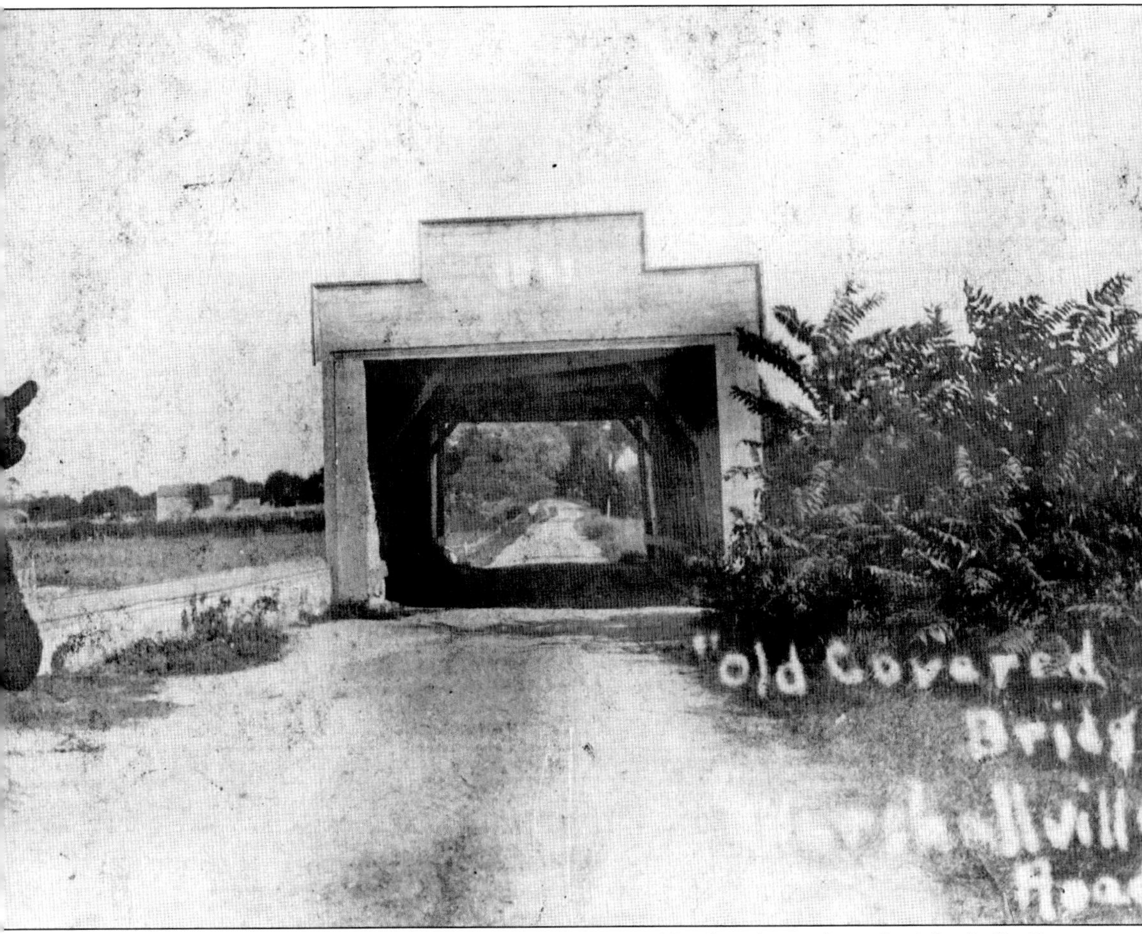

This is an extremely rare photograph of the "old covered bridge" located on the Marshallville Road from Tuckahoe, Cape May County. It was built in 1841 over Mill Creek upstream from the creek's confluence with the Tuckahoe River. Close examination reveals no large truss beams within. Its length suggests a possible Town truss. The bridge shares the name Marshallville along with Tuckahoe. The name Tuckahoe is of Indian origin. The settlement of Tuckahoe began along both sides of the river. The river bustled with activity in the early 19th century with three major industries flourishing: shipbuilding, bog-iron production, and glass-making. Glass-making was thriving in the southern New Jersey area with John Estell Glassworks to the north of Tuckahoe. On the banks of the Tuckahoe, Randolf Marshall owned and operated a glassworks in 1814. The historic town of Marshallville has a large collection of early 19th-century homes, and the original brick and clapboard homes still dot the river's edge. The bridge itself was replaced in 1901.

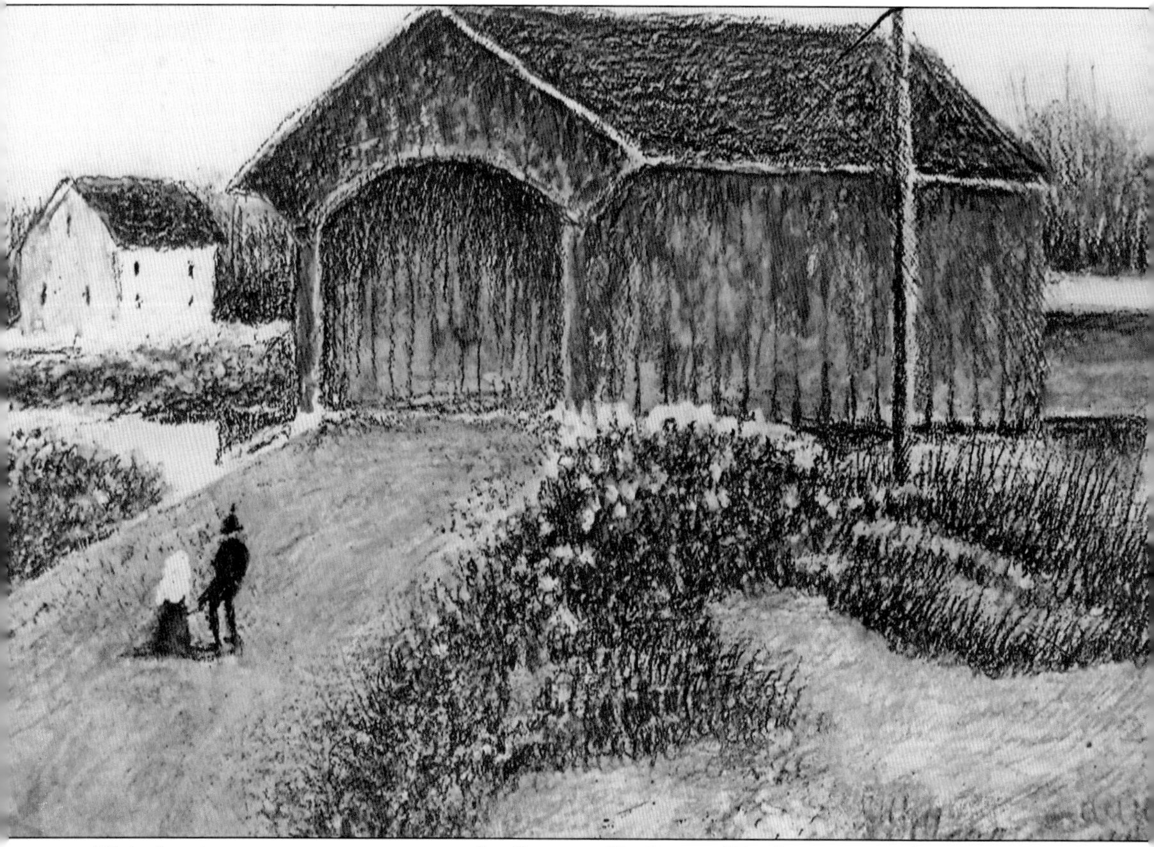

This drawing captures a moment at the Dennisville Covered Bridge crossing the Dennis Creek in Cape May County. The artist was the late Larnie (Lee) Cassidy who had lived in Green Creek and drew this image on wood. Another image of this bridge existed in a wall mural painted by Ray Dixon, who owned the Nathaniel Holmes Jr. House, a part of the Dennisville Historic Home Owners Association. Again, the bridge was shown as a single covered span of unknown truss type with closed sides, as seen by Bob David, a more recent owner of this house. This covered bridge was built by John Wilson of East Creek in 1844. It replaced a bridge built over the creek connecting a causeway in 1789. The *Star of the Cape* reported the bridge was sold at auction in 1885. (Courtesy of Jack Connolly, DHHOA.)

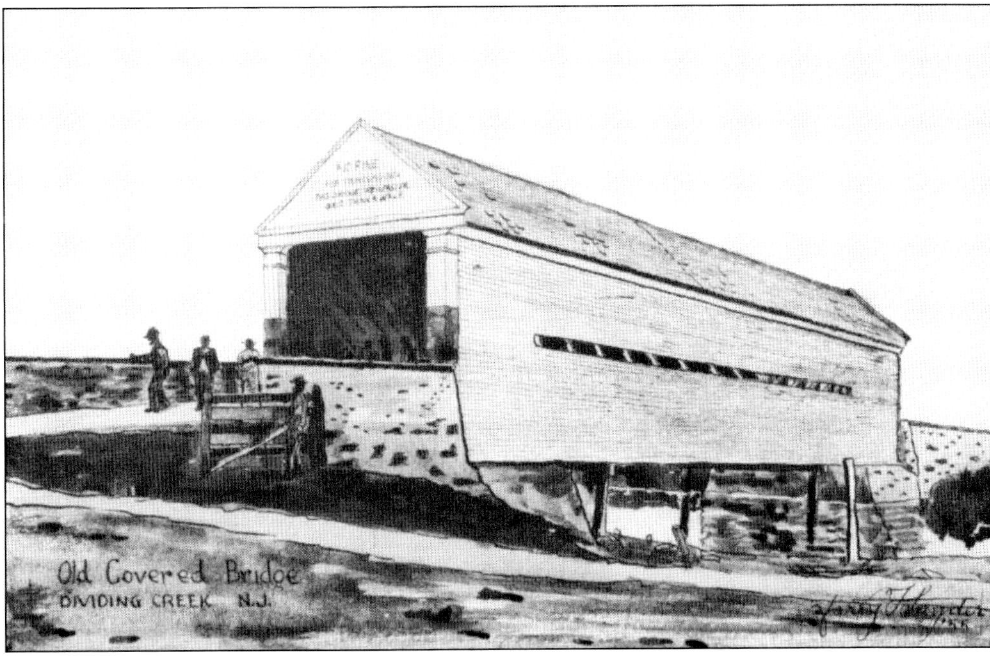

This single-span covered bridge in Dividing Creek, Cumberland County, replaced a bridge built in 1805. The new bridge of 1841 measured 16 feet wide and 60 feet long and was a Town lattice-truss structure. There was a warning sign indicating a $10 fine for traveling over this bridge at a gait faster than a walk, a common restriction for the wooden bridges built in the 19th century. In the late 1800s, however, this bridge was replaced by one made of iron. According to Brydon, an interesting story related to the Dividing Creek Covered Bridge tells of cattle being let loose in the winter months to graze; to seek warmth, the animals would gather inside the covered bridge. An unsuspecting night traveler was bound to stumble upon some surprise waiting in the unlit bridge.

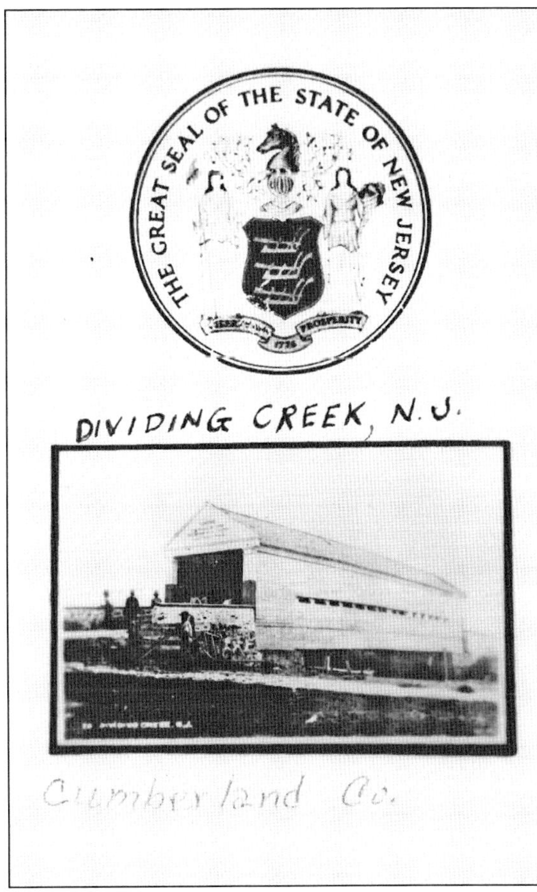

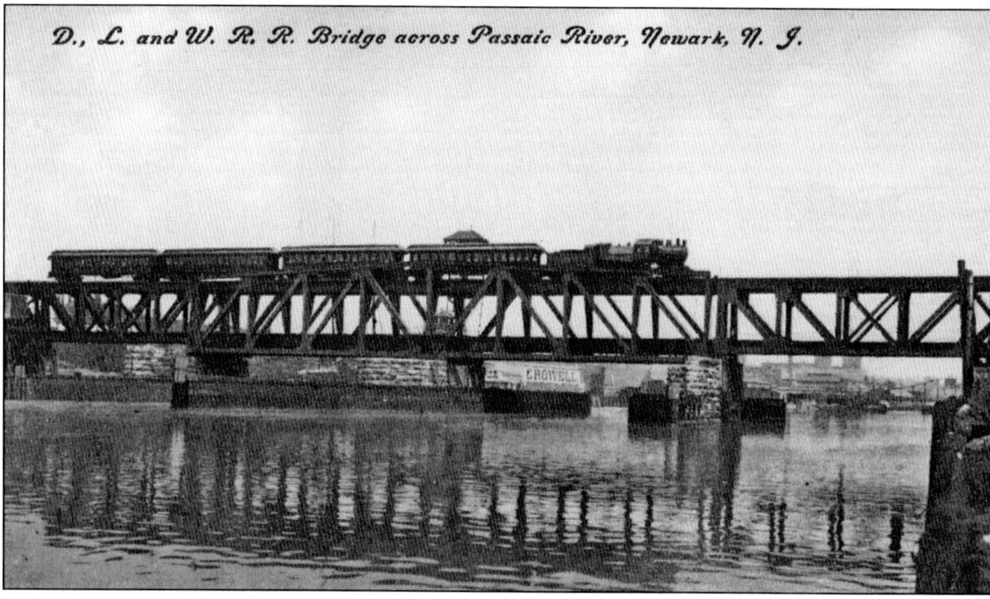

In Essex County, Newark's first span across the Passaic River was a wooden drawbridge at Bridge Street. The bridge above was a center-bearing swing bridge built in 1903. It is an unusual example of a two-level swing bridge designed with two main tracks on the upper level and a single freight track below.

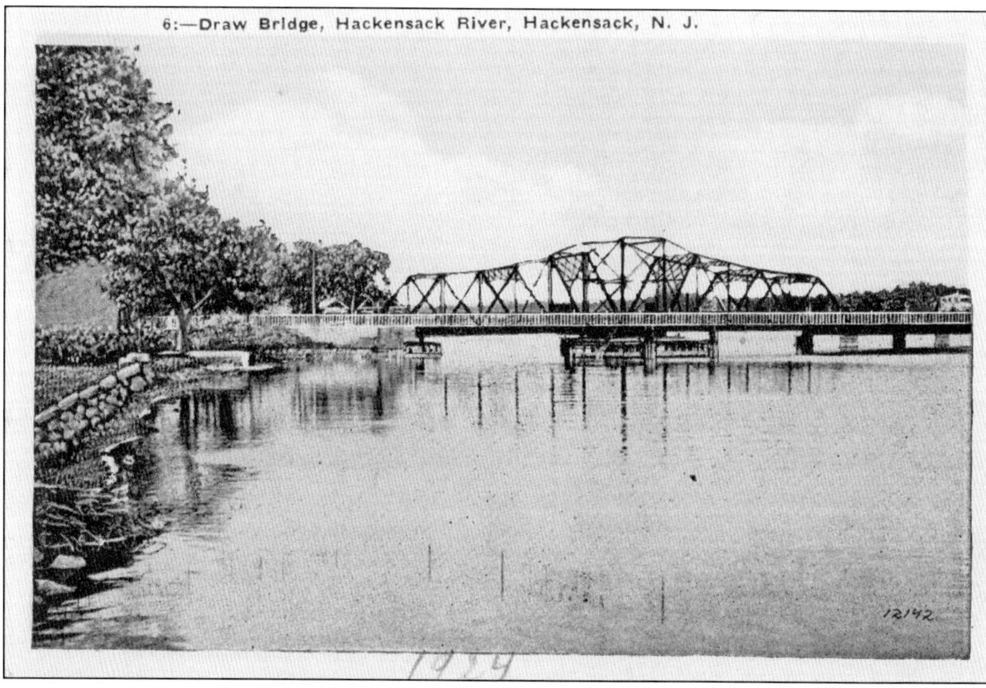

In Bergen County, in the late 1700s, increased river traffic was joined by road traffic, requiring the construction of many drawbridges spanning the Hackensack River. The New York, Lake Erie & Western Railroad built this deck-plate girder with a through-steel-truss swing span around 1905. An operator's shed and concrete interlocking tower were located nearby. (Courtesy of the New Jersey Historical Society.)

Four

The Gateway Region

The Gateway Region of New Jersey includes the most densely populated cities of the state: Newark, the largest city in New Jersey, in Essex County; Jersey City in Hudson County; Paterson in Passaic County; Elizabeth and Linden in Union County; and New Brunswick in Middlesex County. Bergen County boasts one of the earliest and largest shopping malls in the state.

Yet the history of covered bridges in this region is quite scant, indeed. Brydon believed Essex County might have been home to one of the country's very early covered railroad bridges. In 1833, a stockholder circular indicated two bridges over the Passaic and Hackensack Rivers, describing it as having been "built upon piers formed by driving poles into the river bed" and detailing that Town's plan for the bridge would be adopted.

The Passaic and Hackensack Rivers were an engineering challenge for any bridge builder. The Paterson & Hudson River Railroad bridges were built in 1832, using Long-truss construction. Col. Stephen H. Long personally designed the bridges, and well-known Thomas Hassard, a master railroad-bridge builder from Ireland, supervised construction.

In Hudson County, the New Jersey Railroad & Transport Company built a railroad bridge on top of the existing Newark Turnpike Bridge over the Hackensack River in 1833. It burned in 1840. Researched by Thomas Kipphorn, insomuch as the first bridge over the river was a double-barrel covered bridge followed by another covered bridge in 1840, it is possible that the turnpike bridge was also a covered structure. An 1871 relief map shows a wrought-iron bridge.

Finally, on an 1896 relief map of Hackensack (courtesy of the Panoramic Maps Collection of the National Archives) there is what appears to be a six-span, single-track, through-truss covered bridge crossing the Hackensack River. The present steel bridge at the same site is 400 feet long, indicating the possibility of a covered bridge that was likely the same length.

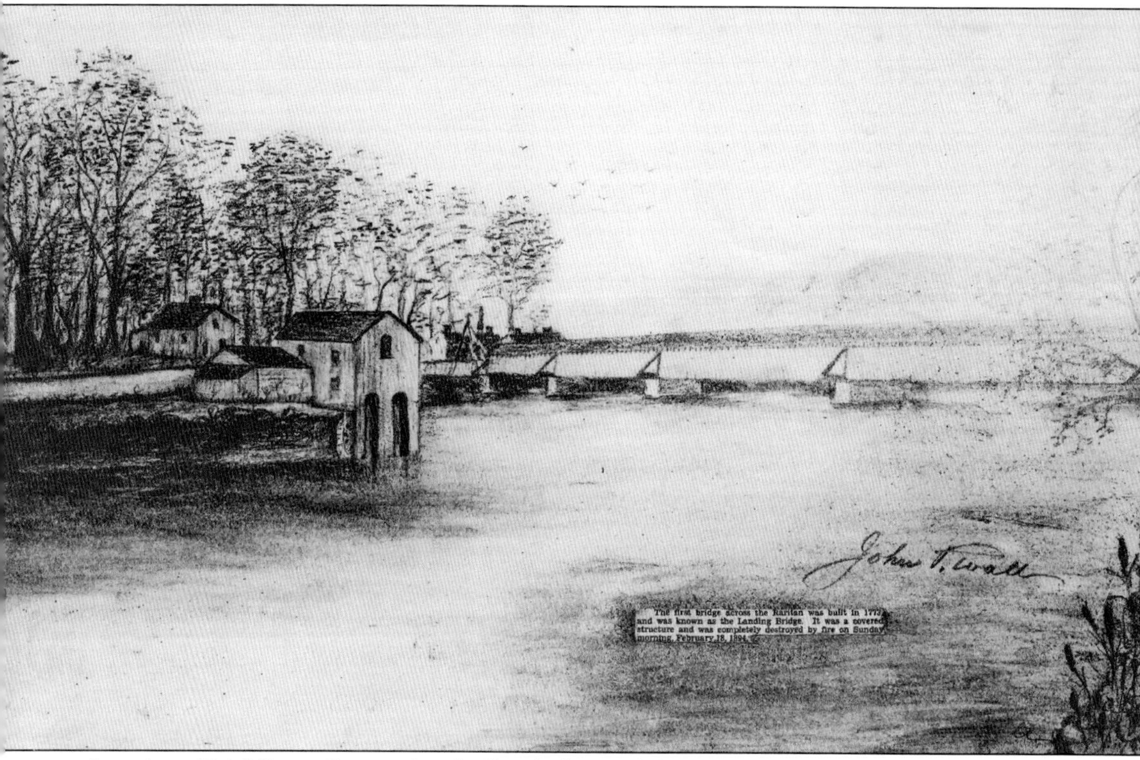

In colonial Middlesex County, local officials directed the construction of roads and bridges, and ferries crossed the Raritan River at several locations. According to *Raritan-Millstone Heritage Alliance of New Jersey Guidebook*, New Jersey's first covered bridge, and one of the earliest in the nation, was situated approximately one mile upstream from New Brunswick. It was erected in 1772 with private subscriptions led by John Duyckinck and Charles Suydam, although the total sum had to be augmented with equal funding from Middlesex and Somerset Counties. In 1776, the bridge was partially burned by the Continental Army in order to slow the British pursuit of George Washington as he retreated from New York to Philadelphia. Lord Cornwallis's stealthy attack, however, prevented the Patriots from completely destroying the bridge. On the second anniversary of the Declaration of Independence in 1778, Washington led his troops along River Road and across the Landing Lane Bridge to a celebration on the New Brunswick side of the river. Finally, in 1895, an iron-and-steel bridge replaced the covered wooden structure. (Courtesy of the New Jersey Historical Society.)

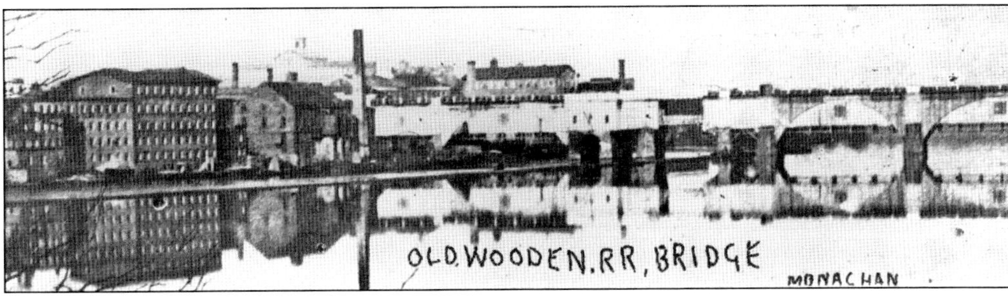

The New Jersey Railroad built a somewhat unusual railroad bridge over the Raritan River in 1837. Again, Col. Stephen Long was involved in the building of this bridge, as was Thomas Hassard. It would seem likely that Long followed his own truss design, which he had patented in 1830. Close inspection of this image clearly reveals arches ending at the piers that supported the bridge. These arches gave the Long truss the extra strength needed for this particular structure. The bridge is estimated to have been 1,700 feet long and made with 17 spans. The two-tiered design had pedestrians and wagons using the enclosed lower level and the trains crossing on the upper level. Fire destroyed the bridge in 1878, and an iron bridge replaced it. (Both, courtesy of the New Brunswick Public Library.)

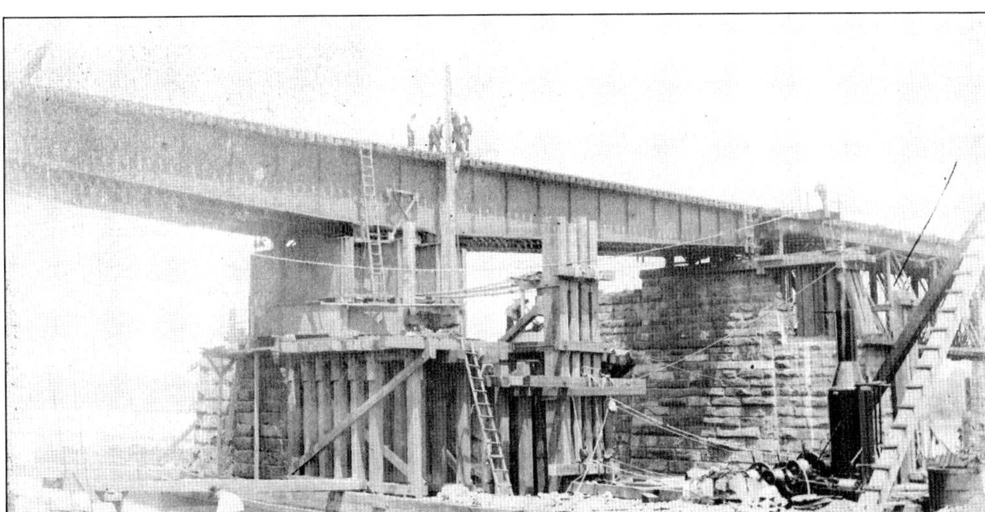

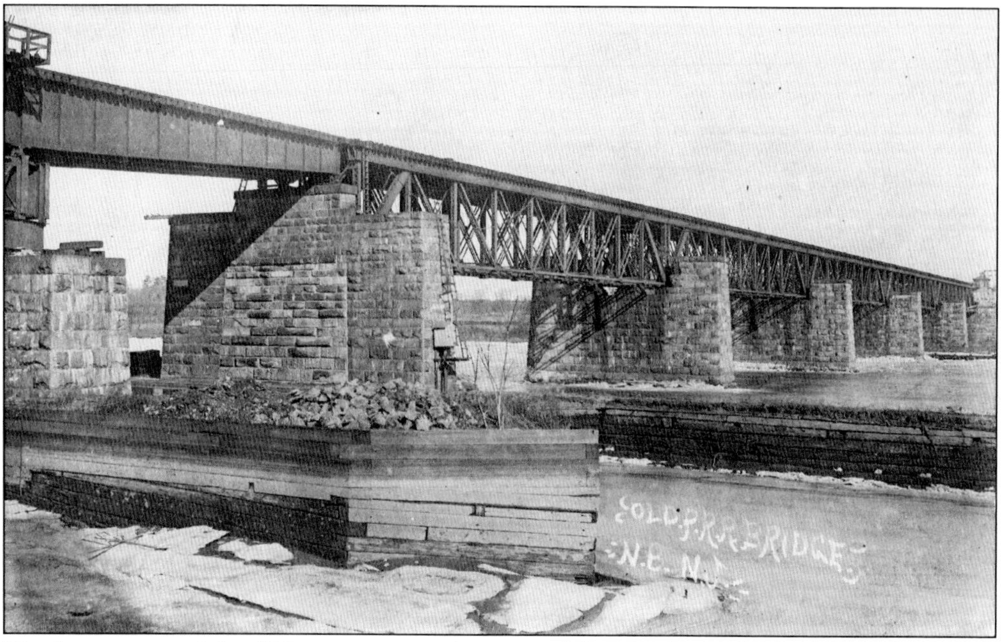

The last quarter of the 19th century was the apex of the metal-truss bridge in America. This image shows the Pennsylvania Railroad Bridge over the Raritan River that replaced the wooden covered bridge in 1878. During this period, bridge designs trended toward greater uniformity and standardization due to advances in the understanding of engineering principles enhanced by experience. In the majority of truss bridges built after 1895, steel replaced brittle cast iron, and because of its higher strength, it also replaced wrought iron. The Pratt-truss designs emerged as the most popular of the myriad truss configurations because of their simplicity. The image below shows the 2,000-ton bridge ready to be moved 14 feet in one operation. (Both, courtesy of the New Brunswick Public Library.)

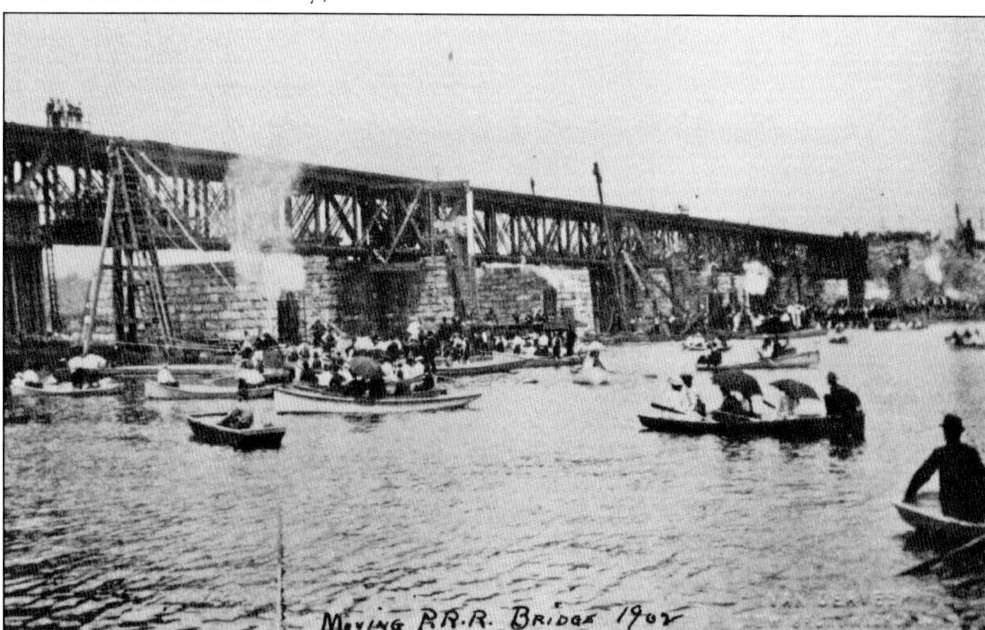

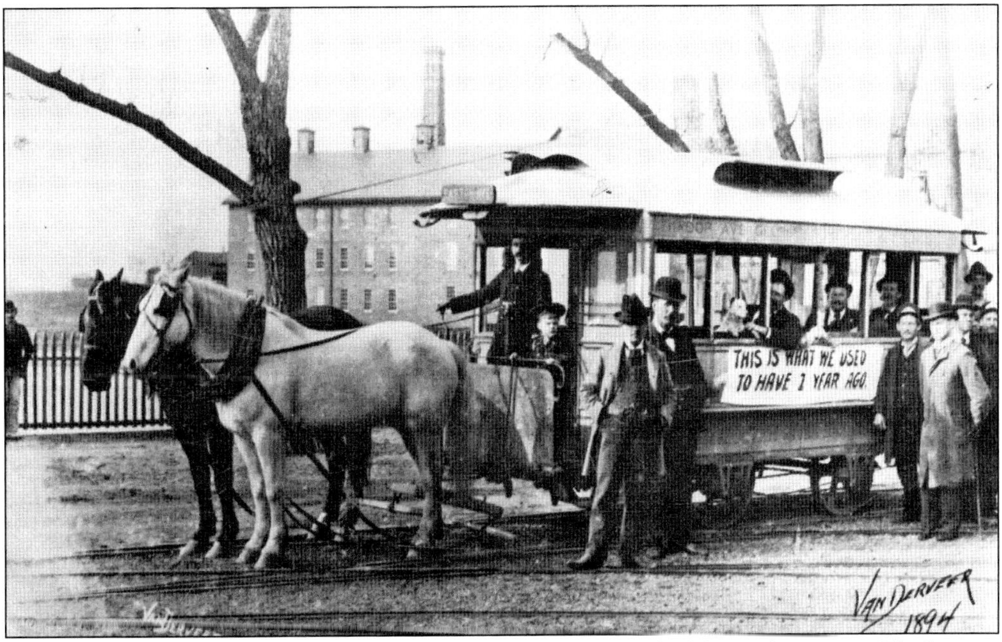

A horse-drawn rail trolley is shown in New Brunswick at the end of the 19th century. The banner reads, "This is what we used to have 1 year ago." The Johnson Brothers came to New Brunswick in 1885 to establish their pioneer gauze and adhesive-tape plant, recruiting many workers from Hungary who used the trolley service. (Courtesy of the New Brunswick Public Library.)

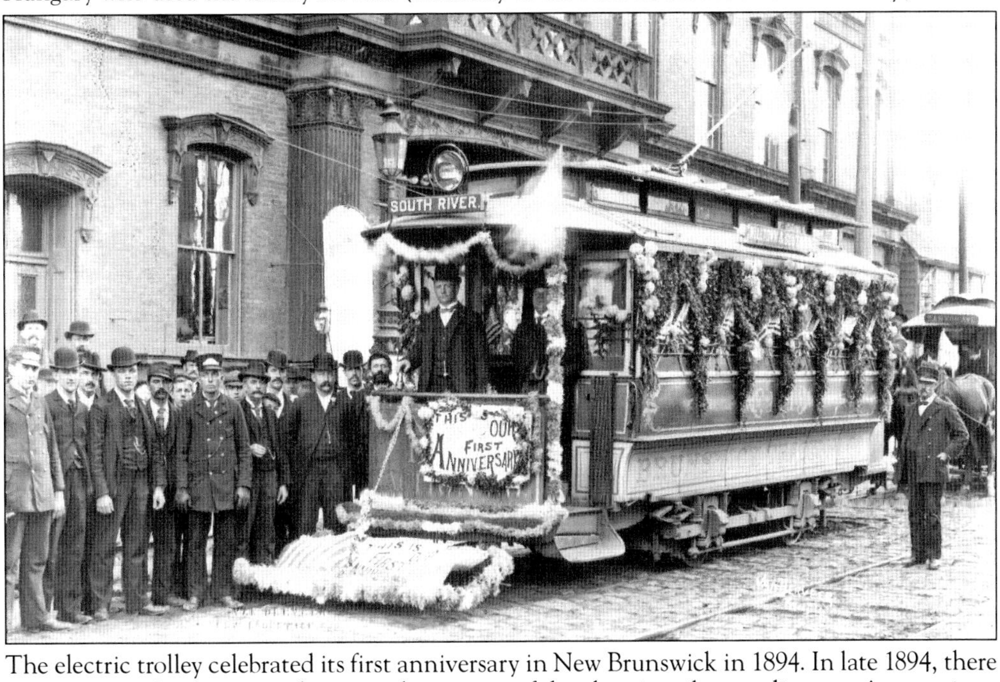

The electric trolley celebrated its first anniversary in New Brunswick in 1894. In late 1894, there were rumors of corruption relating to the passage of the electric railway ordinance. A committee was formed to investigate. Later in 1909, many reports of sightings or encounters with the Jersey Devil throughout the state caused the trolleys in New Brunswick to employ armed guards. (Courtesy of the New Brunswick Public Library.)

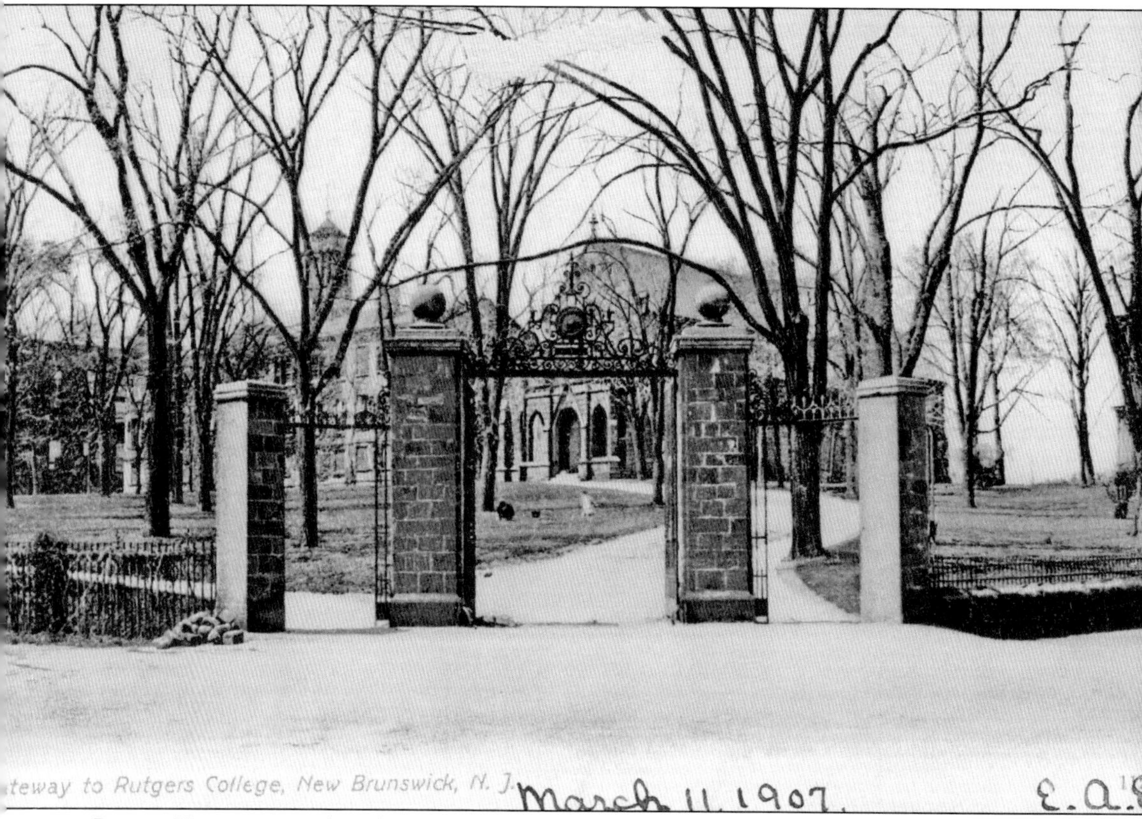

Rutgers University was first chartered in 1766 under the name Queen's College. It was rechartered in 1770. New Brunswick was chosen as the site of the college, which, under the auspices of the Dutch Reformed Church, was originally set up in a former tavern with one faculty member, Frederick Frelinghuysen. The name was changed from Queen's College to Rutgers College in 1825, in honor of philanthropist Col. Henry Rutgers. The college faced collapse when most students enlisted to serve in the Civil War. An Act of Congress established a system of land-grant colleges to train students in the mechanical arts and agriculture in 1864, and Rutgers prevailed to become the state land-grant college. This event led to Rutger's emergence as a modern institution of higher learning, also adopting a non-sectarian policy, offering scholarships, and seeking endowments to support academic efforts. In 1924, Rutgers College officially became Rutgers University. It is now the State University of New Jersey. (Courtesy of New Brunswick Public Library.)

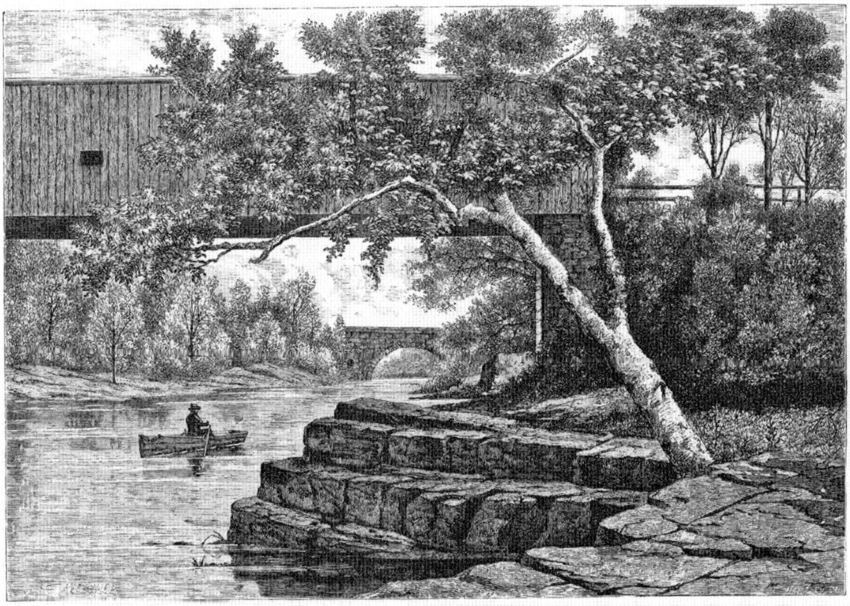

THE PASSAIC, BELOW LITTLE FALLS.

The Little Falls Covered Bridge crossed the Passaic River in Passaic County. The road was then known as Union Avenue. The old Morris Canal, once an important artery of trade and transportation between the Delaware and Hudson Rivers, wound its way through the town. A stone-arch bridge can be seen upriver.

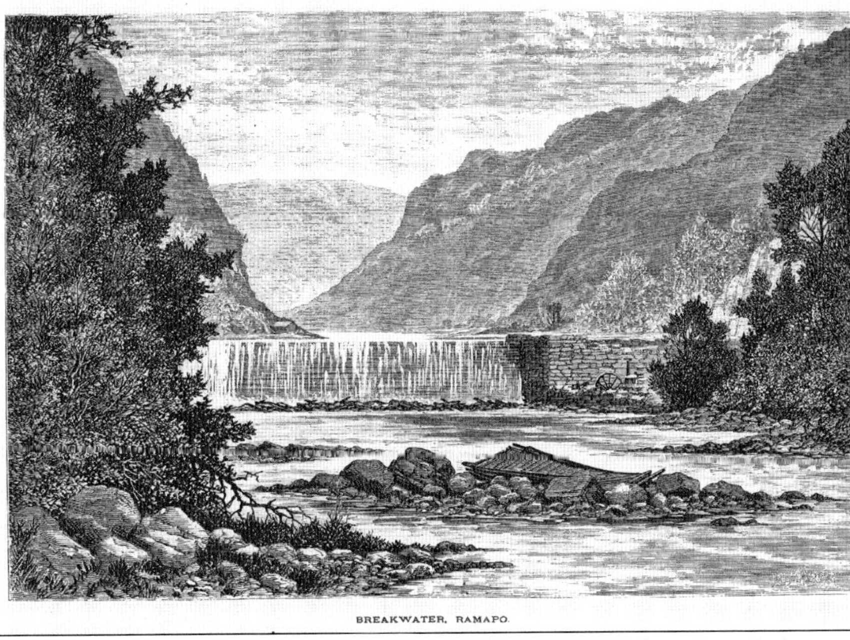

BREAKWATER, RAMAPO.

The Little Falls Covered Bridge was a single span with no arches evident. It might have consisted of a multiple-kingpost or Town lattice truss. The bridge was located near the Beattie Carpet Mill, a major factory in town. The falls under the bridge were blasted out to provide a power source for the mill.

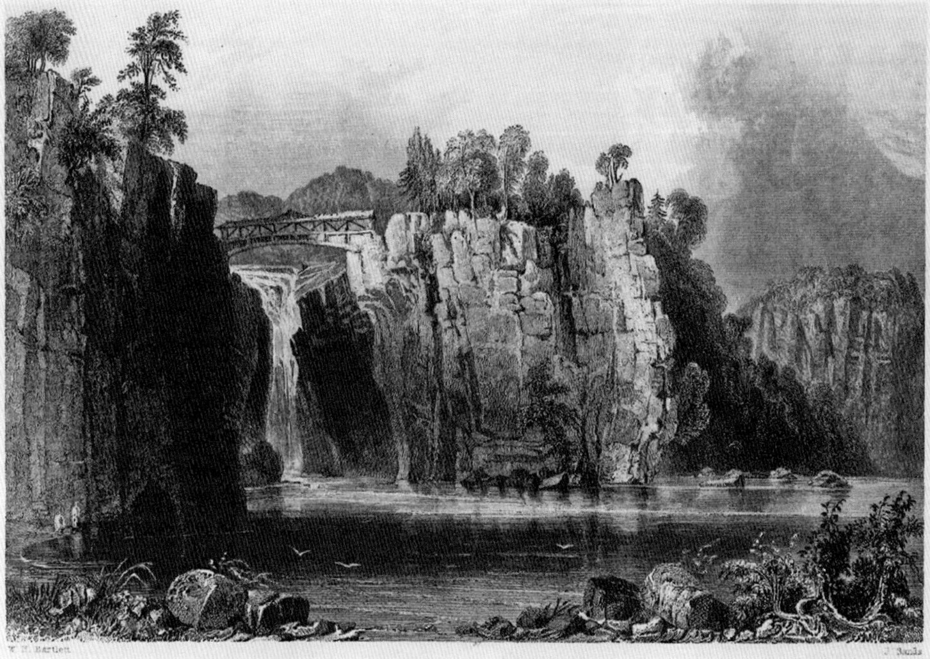

VIEW OF THE PASSAIC FALLS.

The Great Falls at Paterson, Passaic County, have long been known by residents of New Jersey and New York as a beautiful sight worthy of visiting. In the late 18th century, trips by stagecoach brought visitors to the falls. In 1827, Timothy B. Crane bought the north bank of Passaic Falls and turned it into a commercial pleasure garden, called the Forest Garden. The *Paterson Intelligencer* praised it as a retreat where "the refinements of taste and art are combined with the varied and romantic beauties of nature." Historian Paul Johnson wrote, "Crane, however, had to bridge the falls' chasm that separated Paterson from his property. Designed by Crane himself, the covered bridge was made of wood, and its arched bottom sat with heavy grace above the falls. The sides were open, with lattice railings, affording a full view of the falls and chasm." Crane called it Clinton Bridge, after New York governor DeWitt Clinton.

Five

The Skylands Region

The beautiful Skylands Region includes the history of more than a dozen amazing covered bridge structures that connected New Jersey with Pennsylvania over the Delaware River. Hunterdon, Warren, and Sussex are the riverbank counties, with Morris and Somerset completing the region.

In Somerset County, a wooden bridge over the Raritan River was erected as early as 1761 and named the Queen's Bridge in 1767. Later, it became a covered bridge, and during the Revolutionary War, the bridge was used repeatedly by both sides during the Battle of Bound Brook, which was fought on April 13, 1777. Lt. Gen. Lord Charles Cornwallis launched a four-pronged attack on the village of Bound Brook, and about 4,000 British troops defeated the Continental Army.

It is here in Skylands where the only extant covered bridge in New Jersey is found. The federal Historic American Buildings Survey and Historic American Engineering Record programs, established in 1933 to document America's architectural and structural heritage, visited the Sergeantsville Covered Bridge in Hunterdon County in 1937. Up the road from the bridge are the Sergeantsville General Store and the Sergeantsville Inn. Formerly a railroad bridge constructed by the Delaware, Lackawanna & Western Railroad in 1871, the Darlington Bridge at Delaware Station, Warren County, was a metal highway bridge over the Delaware River. The original bridge was a wooden covered bridge built in 1855. Having purchased the bridge for $5,000 in 1914, the Reverend Henry Darlington sold the bridge 18 years later to the Joint Bridge Commission for $275,000. In 1836, the first bridge was built at Dingmans Ferry and remained an excellent structure until it washed away in 1847. Over 200 pigeons roosting in the bridge were also washed away and perished. Carol A. Phillips, secretary-treasurer of the Dingmans Choice and Delaware Bridge Company, writes, "It is likely that this first bridge was covered since pigeons typically roost under cover." A second bridge also must have been covered, for in the diaries of old Sussex County residents, there are accounts of hauling snow to the bridge so that sleds could cross in the winter.

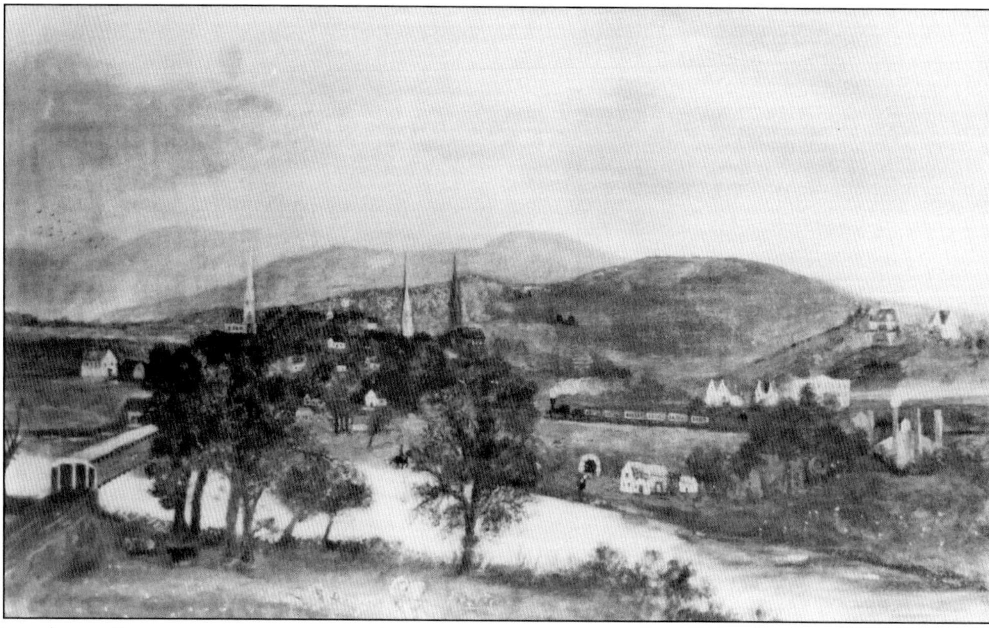

Lambertville was named as a result of the US Post Office Department creating a village post office there. The person petitioning for this was a US senator during the time of Thomas Jefferson. Consequently, John Lambert named the village Lambertville. The original spelling was actually Lambert's Ville, but the *s* was dropped when the town became incorporated. This 1851 painting was created by James Doolittle. (Courtesy of the New Jersey Historical Society.)

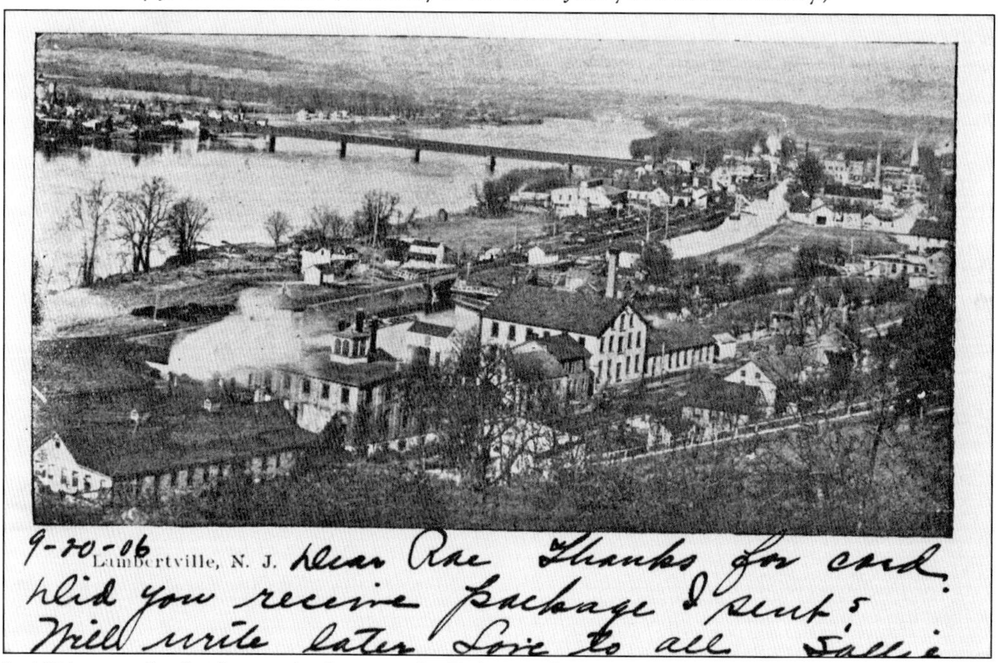

In 1814, a wooden bridge was built across the Delaware River, connecting Lambertville, Hunterdon County, with the town of New Hope, Pennsylvania. Two businessmen from New Hope—Benjamin Parry and Samuel Ingham—received legislative permission to build a wooden covered bridge with six spans. Construction costs included the bridge, tollbooths, and purchasing the ferry rights.

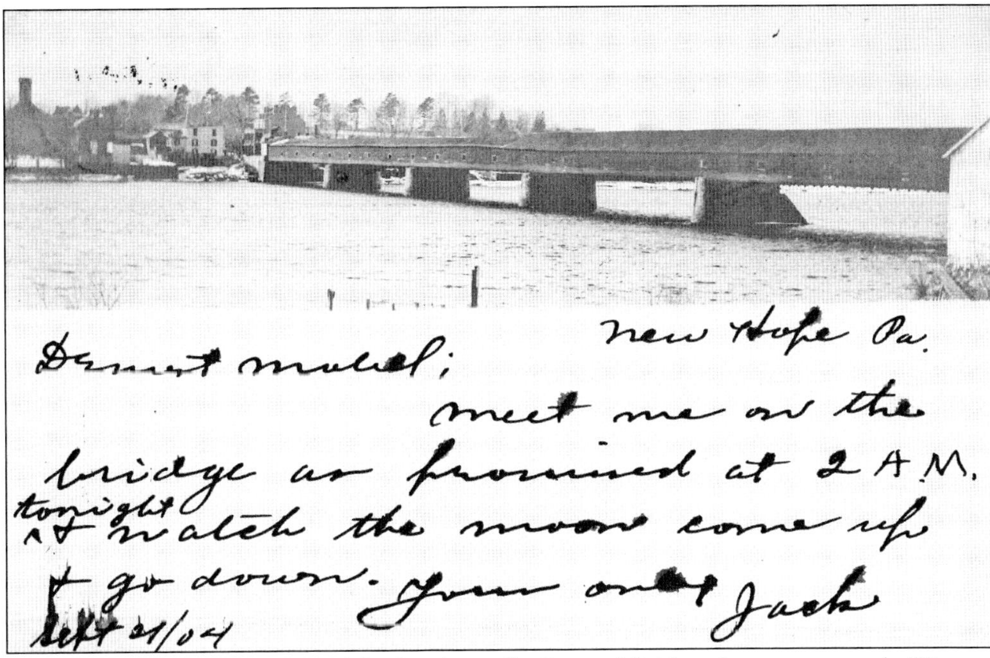

On an old real-photo postcard of the Lambertville Covered Bridge, the message reads, "Dearest Mabel; meet me on the bridge as promised at 2 AM tonight and watch the moon come up and go down. Yours only, Jack." As is often written, a quiet covered bridge at night was not only a venue for transportation.

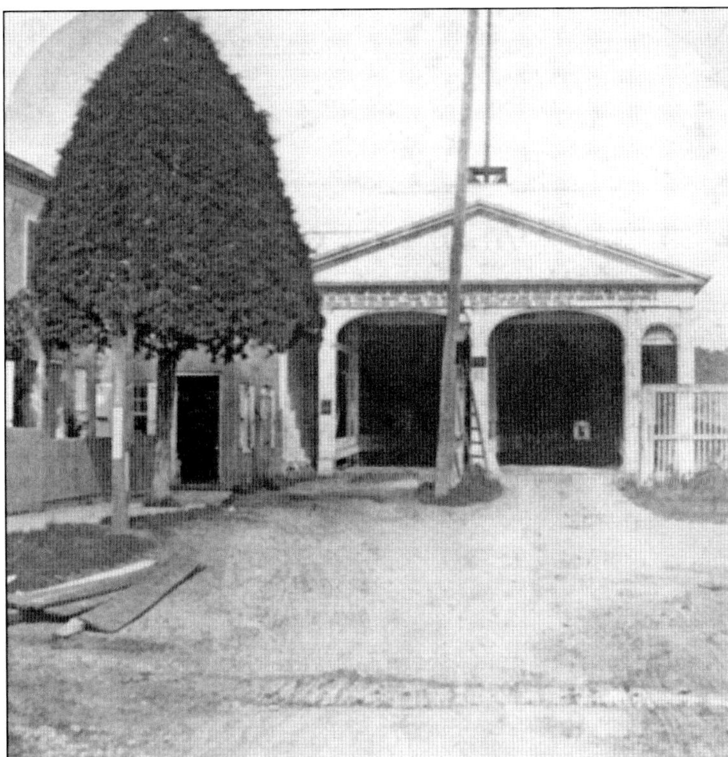

The portal entrance to the covered bridge on the Lambertville side of the Delaware River appears to be less cluttered than the New Hope side. Tolls were collected in New Jersey. One restriction declared that no more than 10 cattle could be driven across the bridge at one time. (Courtesy of Todd Clark.)

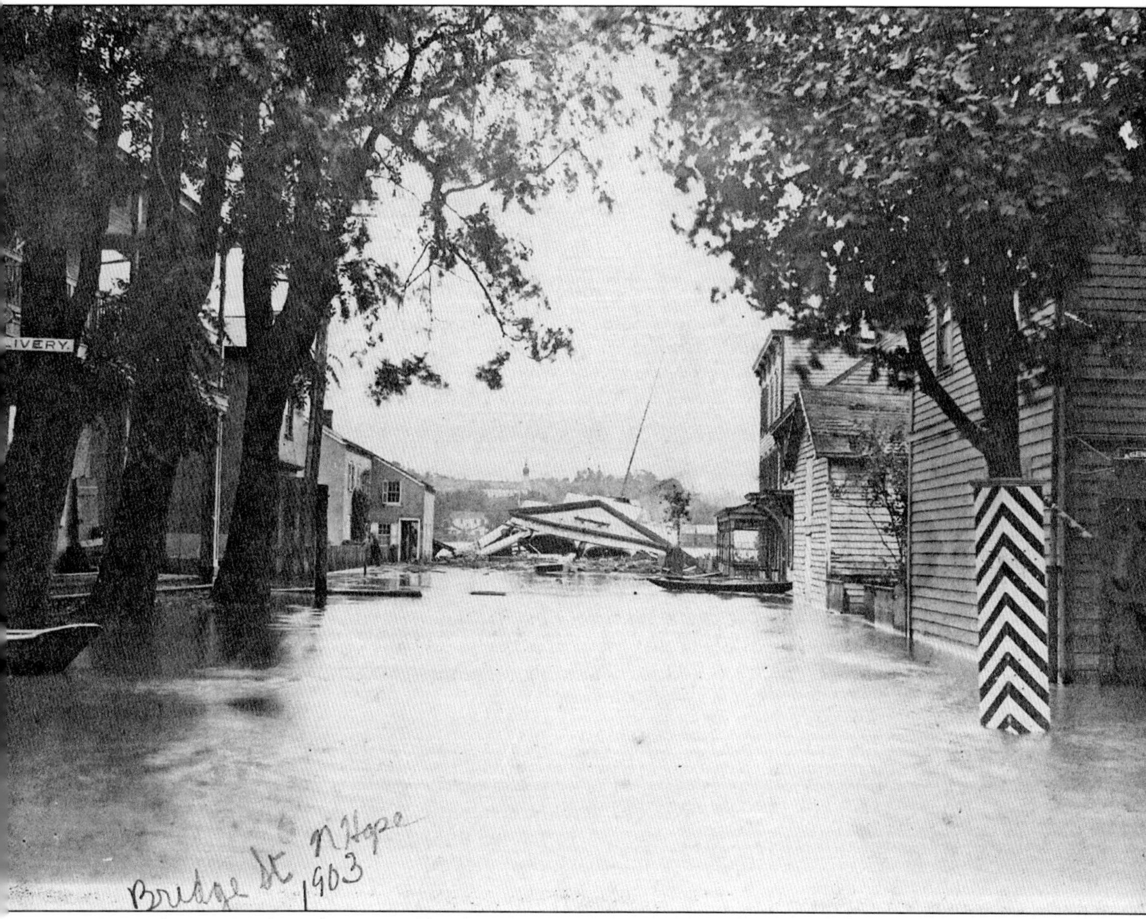

According to Brydon, the covered bridge that connected Lambertville with New Hope, Pennsylvania, was the longest of the Delaware River bridges, at 1,051 feet long with six arched spans, each 175 feet in length. It was a double-barrel (two-lane) bridge with a walkway. Rebuilding sections had been necessary in 1841 and 1842. Following the catastrophic flood of 1903, this is all that remained of the bridge and its portal at New Hope. Flooding is evident far beyond the banks of the river. The arches of each span of Lewis Wernwag's design had rested atop the abutments and piers, which were an innovation in covered bridge design. This would be the last wooden covered bridge to link Lambertville to New Hope. (Courtesy of the Lambertville Historical Society.)

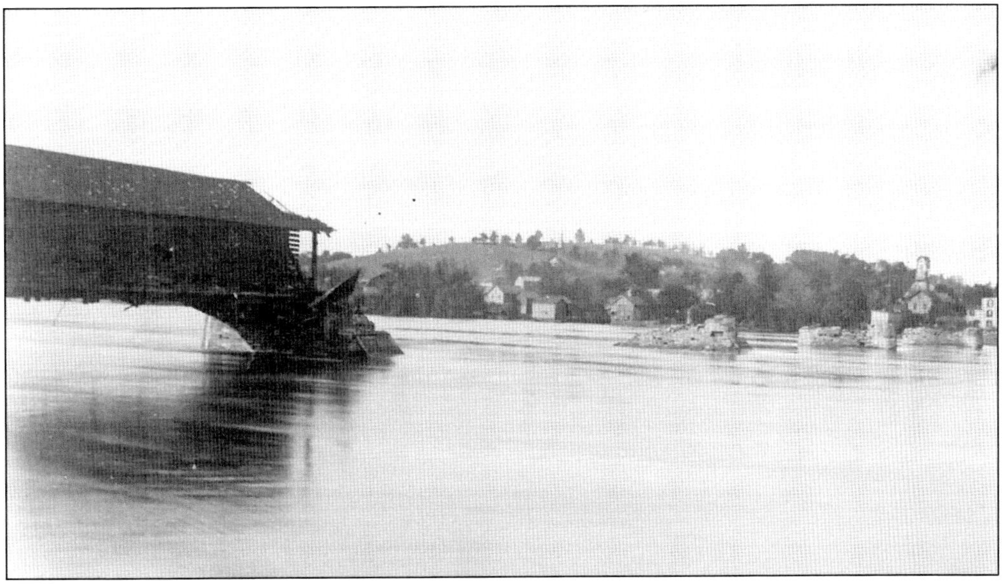

The Delaware is the longest undammed river east of the Mississippi, extending 330 miles from the Catskill Mountains of New York to the mouth of the Delaware Bay, where it meets the Atlantic Ocean. In October 1903, the rain nearly washed out the first-ever World Series and grounded the Wright brothers' glider in Kitty Hawk, North Carolina. While along the Delaware River Valley, it was an utter disaster. Helpless, residents of towns like Lambertville watched as the river rose relentlessly, its usual calm and shallow waters spilling over its banks, reaching levels of up to 28 feet. According to Jon Blackwell in the *Trentonian*, the torrent swept away everything in its path. (Both, courtesy of the Lambertville Historical Society.)

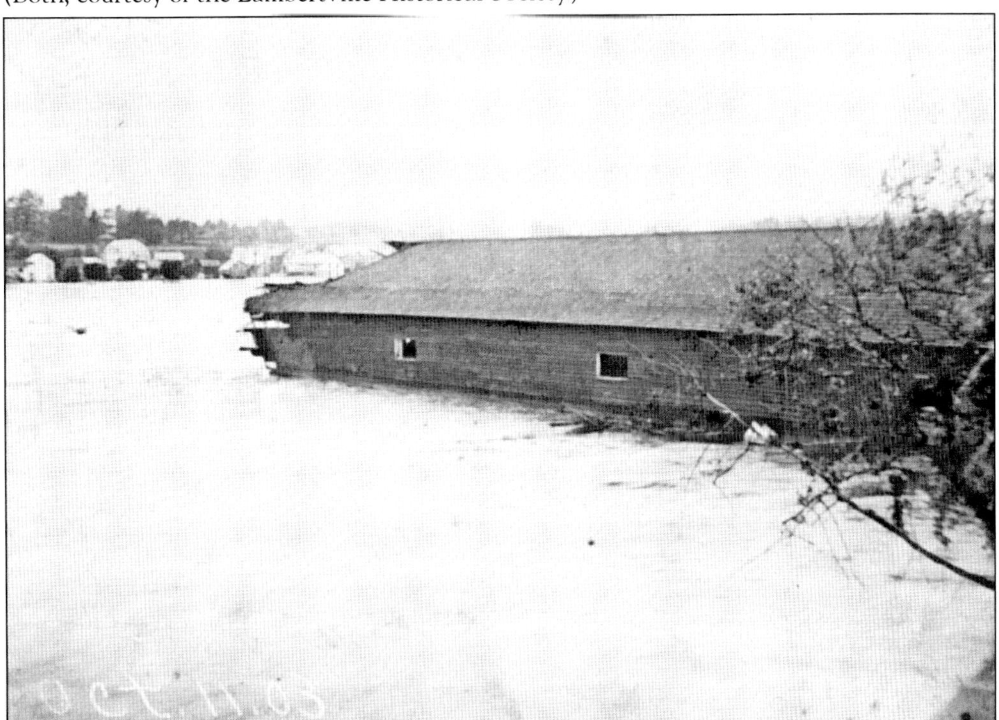

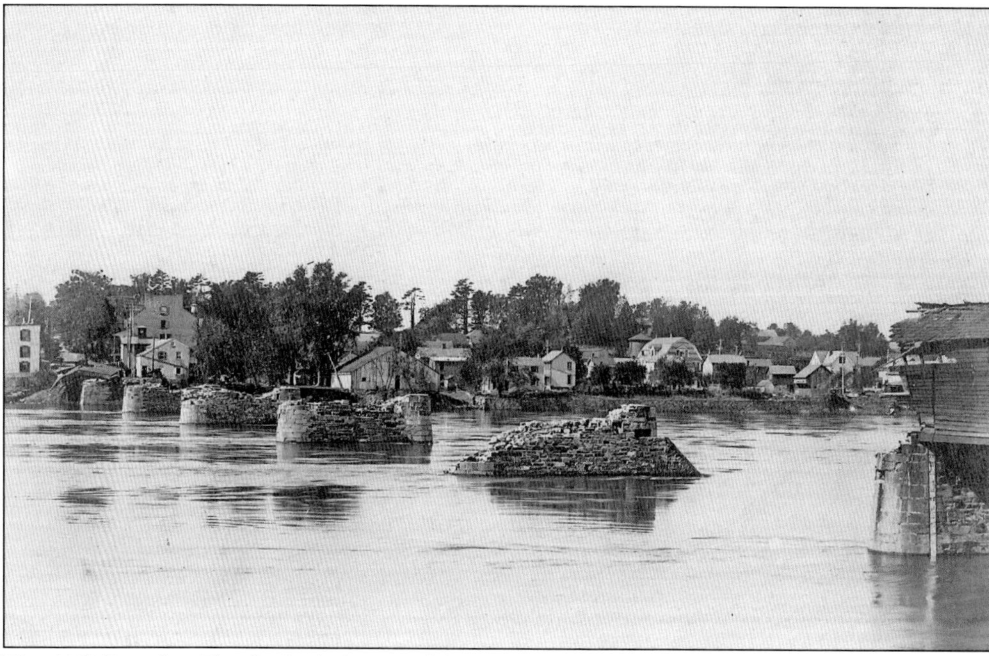

The 1903 Pumpkin Flood, which destroyed the covered bridge at Lambertville, began on October 7 as a tropical storm along the Virginia Capes, as reported by Jon Blackwell in the *Trentonian*. On that Saturday afternoon, the rain resumed, and an immense surge of water came pouring into the Delaware Valley from the Lehigh River. (Courtesy of the Lambertville Historical Society.)

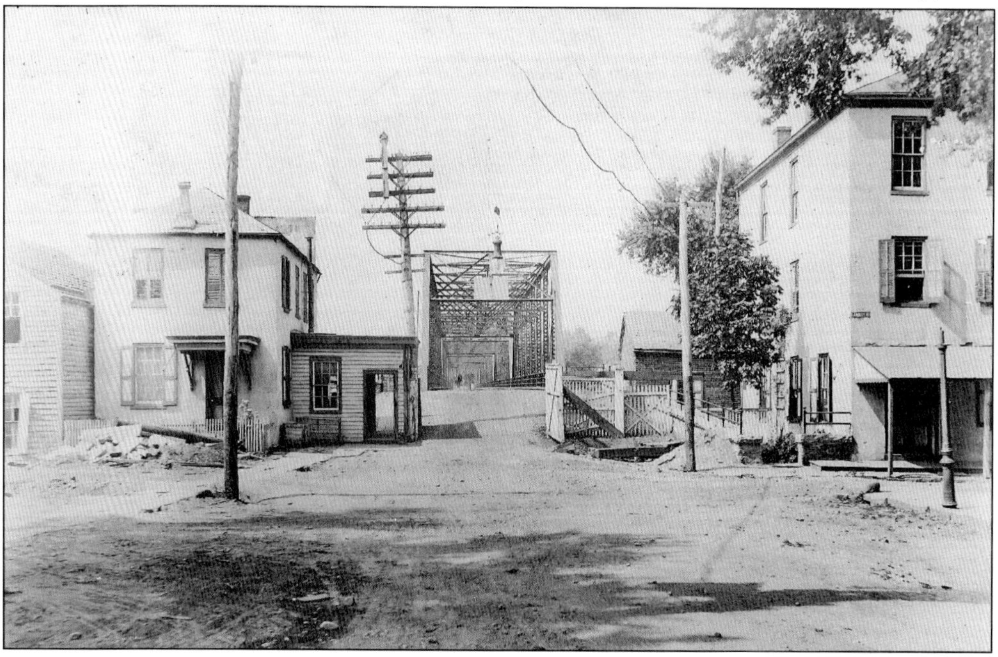

The present six-span steel truss took the place of the wooden covered bridge following its destruction in October 1903. The new bridge was built on the original abutments and piers, which did need to be repaired. Still a company bridge in 1905, tolls continued to be collected. (Courtesy of the Lambertville Historical Society.)

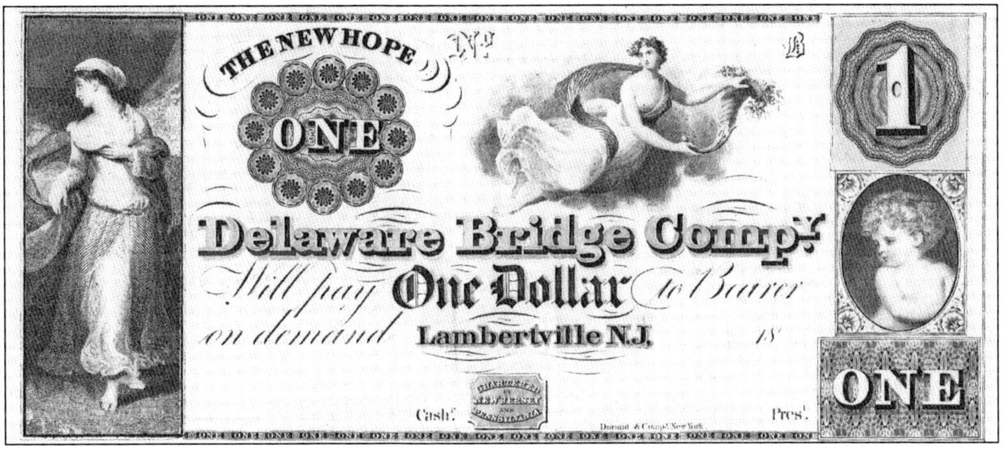

In the United States, the Free Banking Era lasted from 1837 to 1863. Almost all places could issue their own paper money: states, municipalities, railroad and construction companies, stores, restaurants, and even churches. The early Delaware Bridge Company was allowed to serve the area as the only local bank, and the owners paid more attention to the interest rate they could charge than to the efficiency of their bridge. The company issued paper notes from 1836 through 1841. Bridge tolls were often used illegally for outside investments rather than the maintenance of the bridge. The financial losses of which were greatly increased by the cost of damage repair in the major flood of January 1841. The National Bank Act of 1863 officially ended the Free Banking Era.

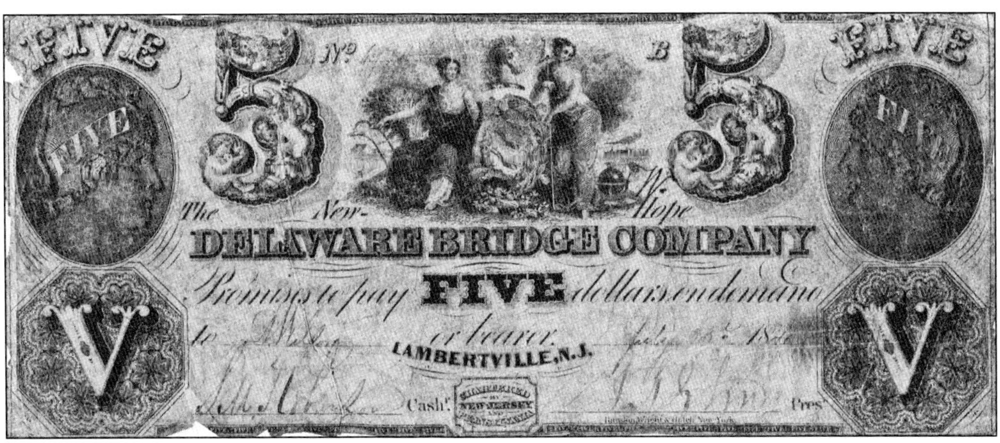

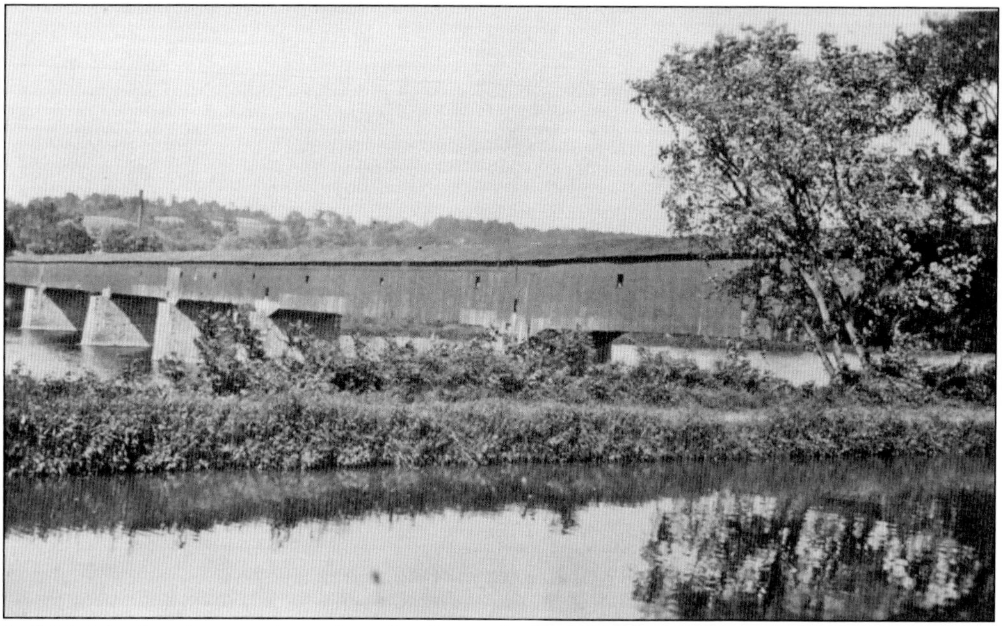

In 1703, John Reading purchased land bordering the Delaware River from local Native Americans. He established a ferry, and the area was called Reading's Ferry. By 1792, John Prall had built a gristmill, sawmill, oil mill, and store. A covered bridge was completed in 1814, with six spans. Called Centre Bridge, the surrounding area was quite active. A canal was finished in 1834, causing population to increase, but in 1841, a serious flood broke the covered bridge in two, at which time it floated down the river and struck the Lambertville Covered Bridge. In 1850, hotel owners exchanged property with the Centre Bridge Company to make a road in a straight line from the bridge to the front of the hotels on Bridge Street. (Above, courtesy of Hunterdon County Historical Society; below, author's collection.)

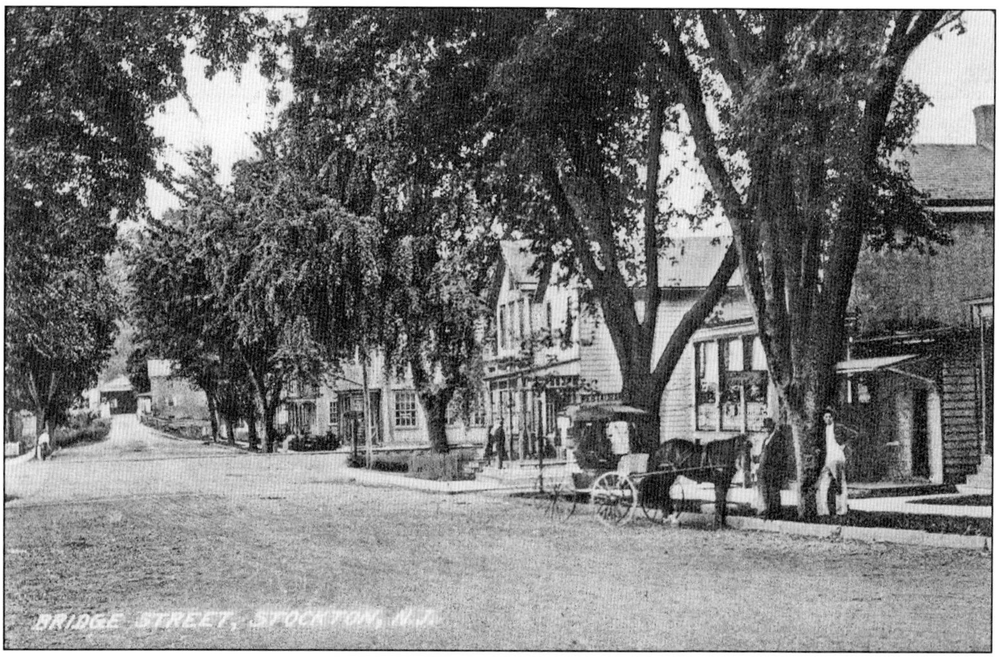

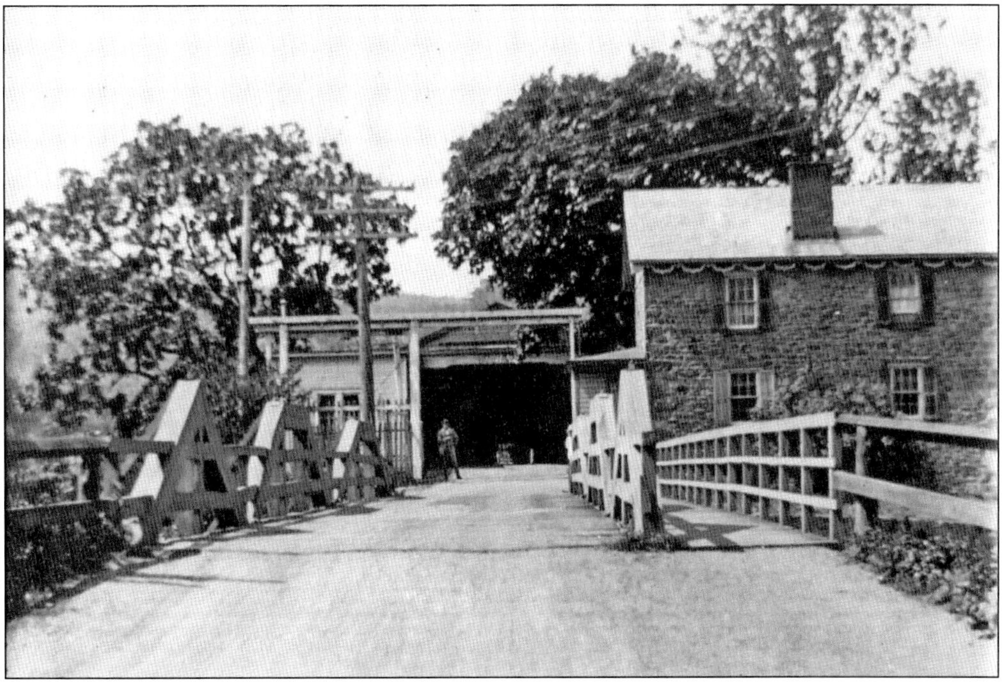

The Centre Bridge, also called the Stockton Bridge, was one of the oldest covered bridges on the Delaware. About halfway between the bridges at Trenton and Phillipsburg, it connected Centre Bridge, Pennsylvania, with Stockton, Hunterdon County. Tolls were paid on the New Jersey side, with a comfortable stone house provided for the toll collector. (Courtesy of the Hunterdon County Historical Society.)

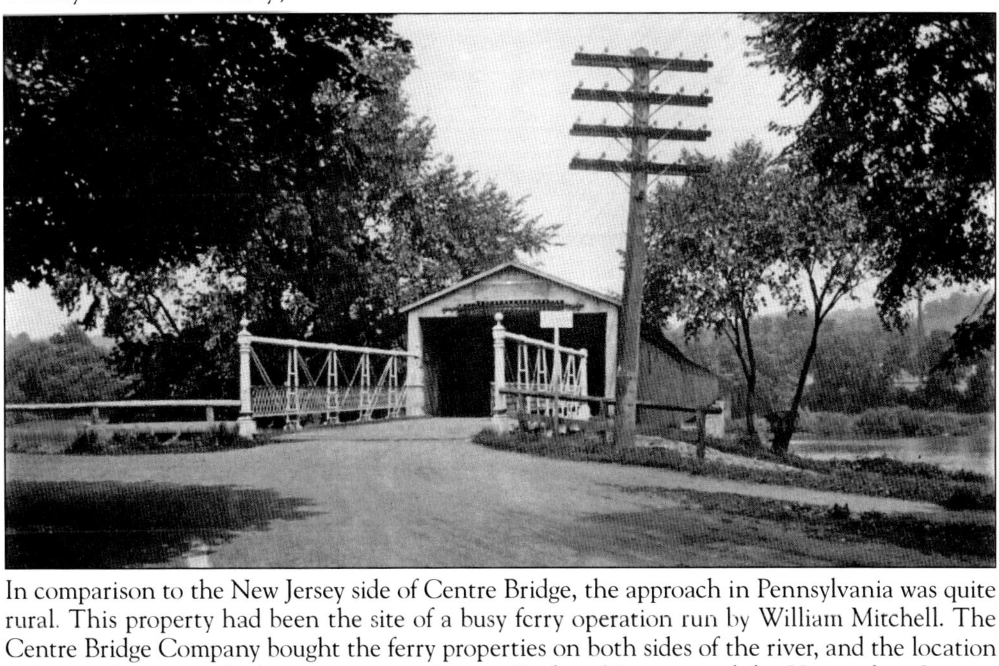

In comparison to the New Jersey side of Centre Bridge, the approach in Pennsylvania was quite rural. This property had been the site of a busy ferry operation run by William Mitchell. The Centre Bridge Company bought the ferry properties on both sides of the river, and the location in Pennsylvania took the same name, Centre Bridge. (Courtesy of the Hunterdon County Historical Society.)

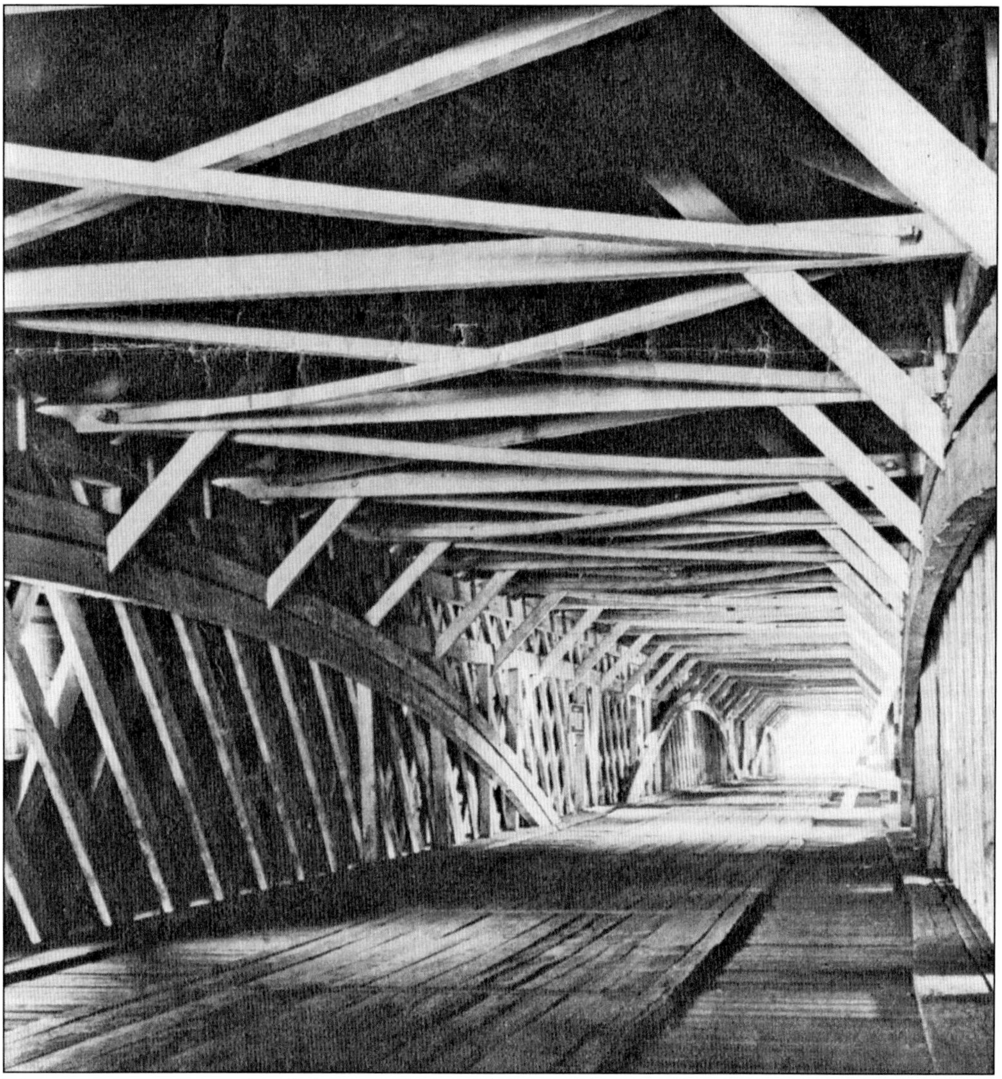

The caption on this undated photograph republished in the *Herald Tribune* reads, "This Wooden Covered Bridge . . . Spanning the Delaware at Stockton. Unadorned except for posters heralding the wonders of the coming circus and distinguished for its utilitarian simplicity, this relic of the horse and buggy days is one of the few remaining in New Jersey." Pelig Kingsley and Benjamin Lord built the Centre Bridge in 1814, the generally accepted date of construction. At a length of 821 feet, the bridge had six spans. The photograph illustrates a wonderful example of Town lattice and arch construction, but this was the result of at least two other builders, including Amos Campbell and Courtland Yardley, who had to make major repairs to the bridge in 1830 and 1841. Twice, three spans were lost to floods. Lateral bracing is also evident to help stabilize the long bridge from the winds swirling down the river. (Courtesy of the New Jersey Historical Society.)

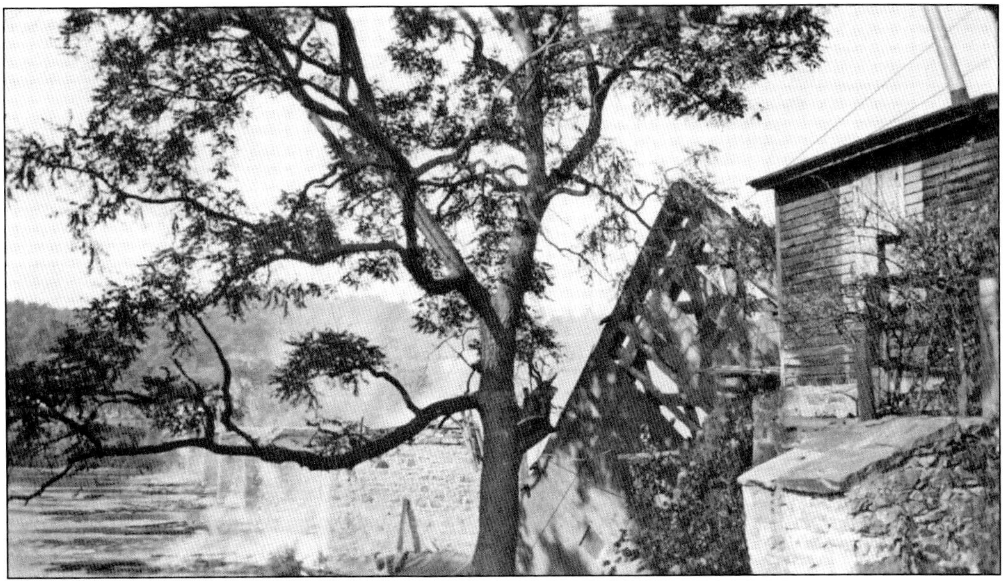

On July 22, 1923, lightning struck the wooden covered bridge crossing the Delaware River at Stockton. The old structure had caught fire twice in the spring of the same year, but the damage had been repaired. This time, the bridge was almost totally destroyed. The wrecked first span shown above was located on the New Jersey end. During the fire, this span broke off and fell into the river. Eight members of the Stockton Fire Company and eight volunteers went down with it, plunging 30 feet to the riverbed below. Fortunately, the river was low and there was no loss of life, although several people were painfully injured. After four years, the Joint Free Bridge Commission opened a new toll-free steel bridge. (Above, courtesy of Hunterdon County Historical Society; below, author's collection.)

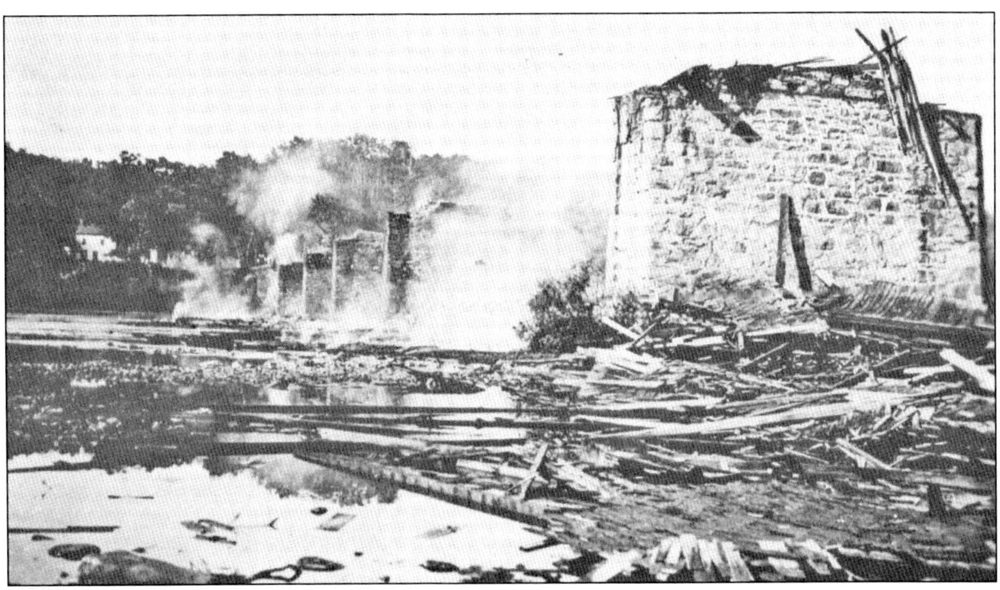

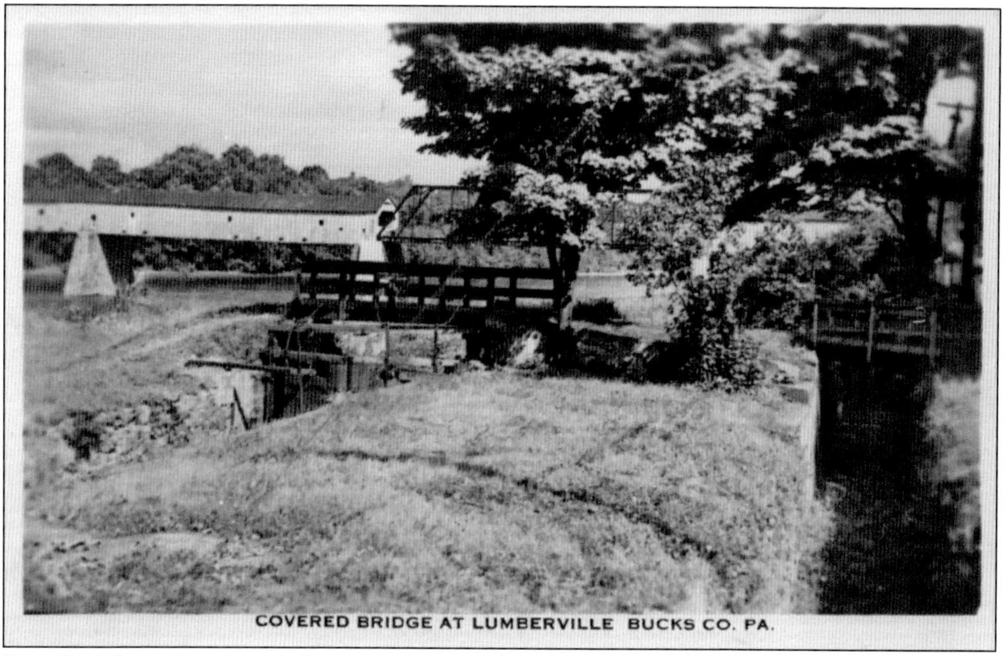

State legislatures approved the construction of a covered bridge crossing the Delaware River between Raven Rock, Hunterdon County, and Lumberville, Pennsylvania, between 1835 and 1836. A lack of financing held up the project until 1853, a delay that actually prevented potential damage from the major river flood of 1841.

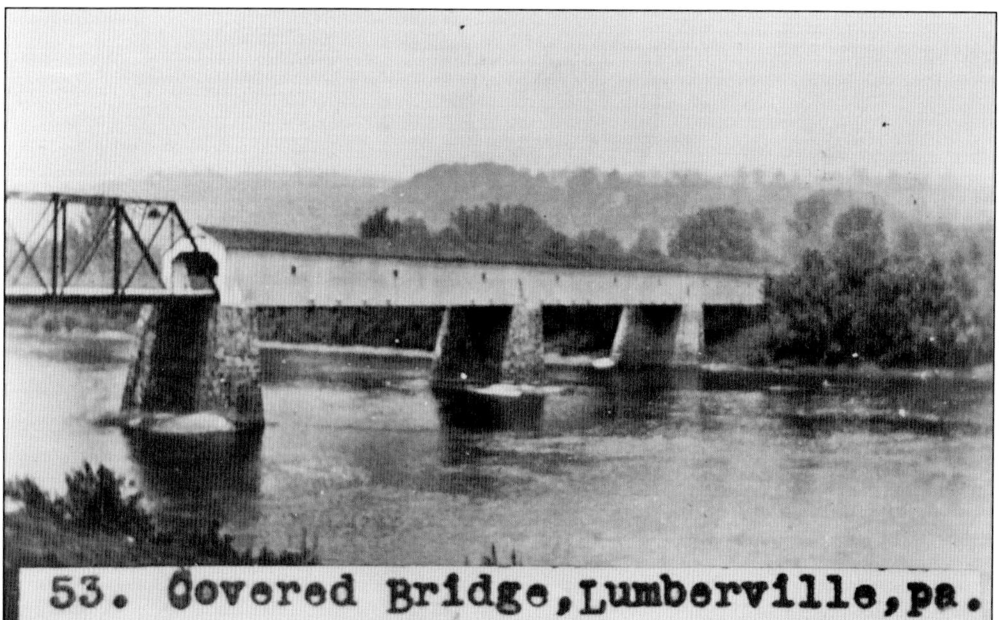

A covered wooden structure engineered by Solon Chapin and Anthony Fry, this four-span bridge was built by the Lumberville Delaware River Bridge Company and opened in 1856. A fifth span crossed the Delaware Division Canal. The area surrounding the bridge never grew to a populated area, so the bridge was not overworked and did not require much maintenance or repair.

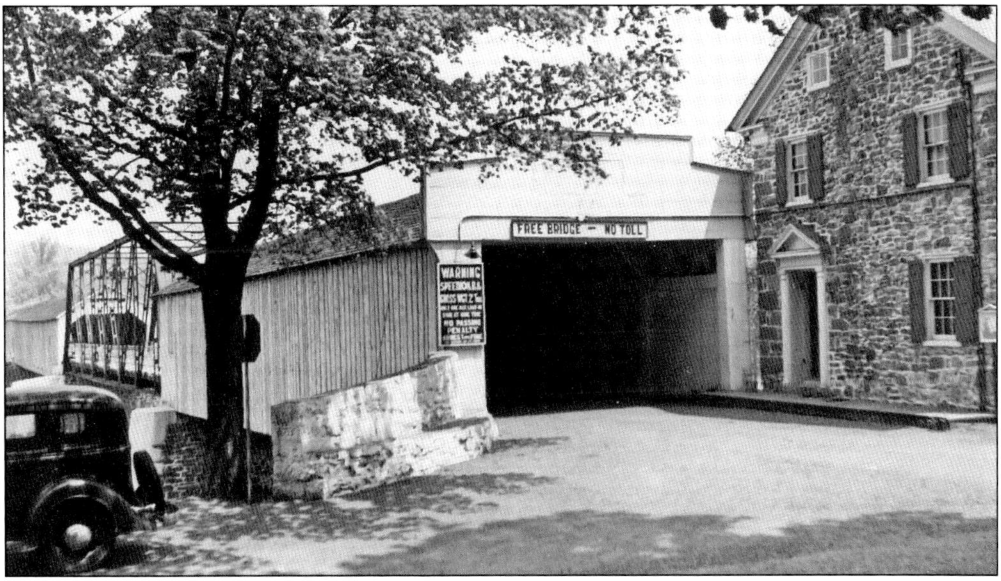

The reported cost for the construction of the covered bridge at Raven Rock to Lumberville was $18,000. The bridge was built in the Town-truss design, at just over 700 feet in length. As with all the Delaware River crossings, this was a toll bridge. Tolls were collected on the Lumberville side, where the toll collector also had a house near the bridge. The photograph above shows the bridge entrance with a new sign: "Free Bridge—No Toll." This occurred after the Delaware River Joint Toll Bridge Commission purchased the bridge as well as the stone tollhouse on the right. The photograph below offers a view of the extra span that crossed the Delaware Canal alongside the river.

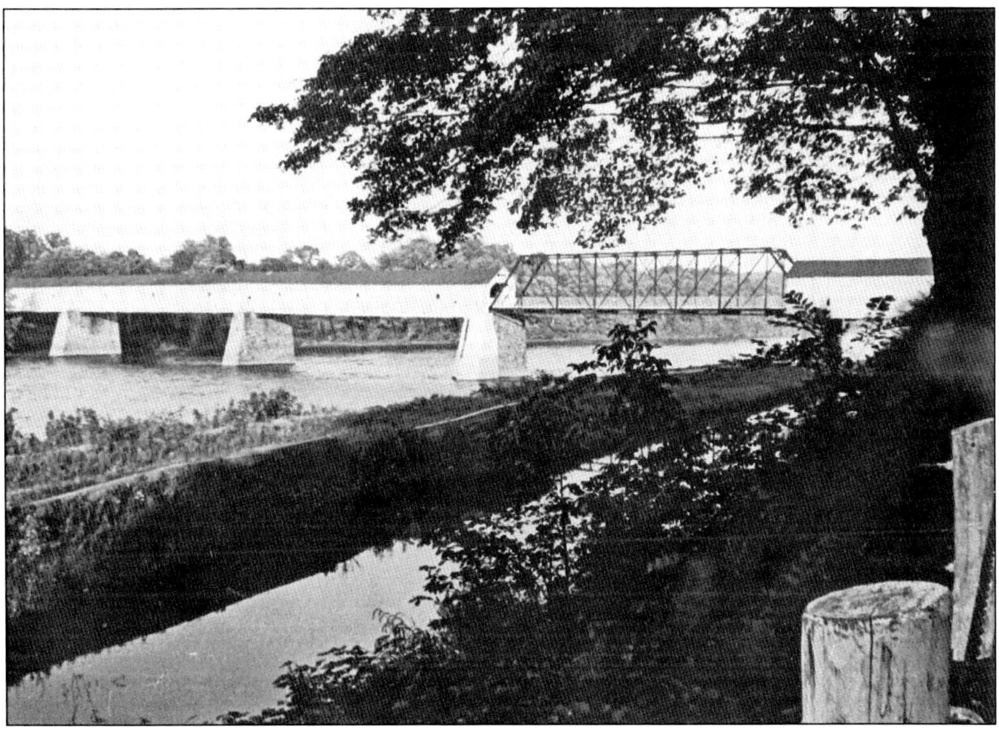

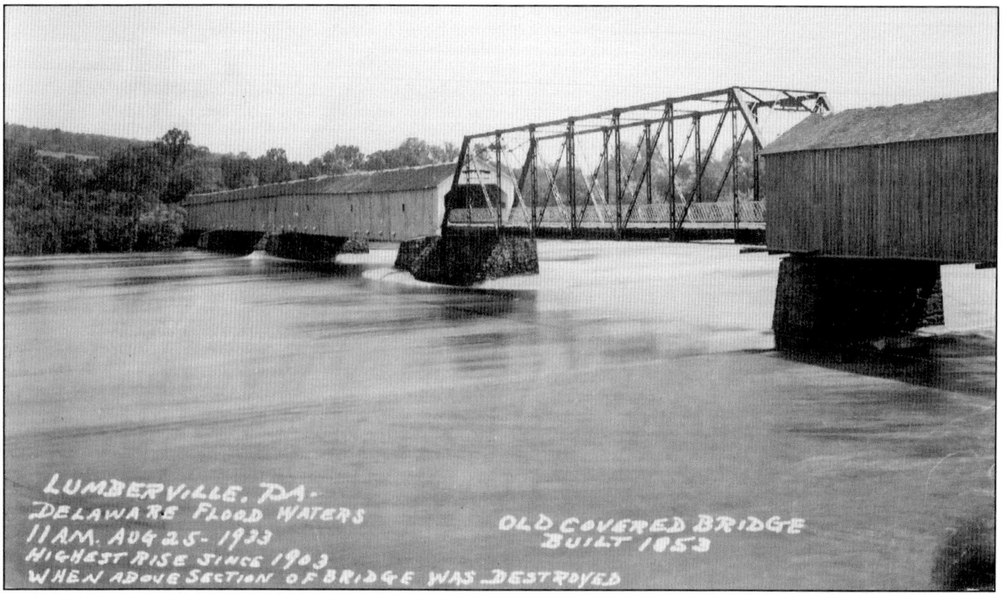

This photograph shows the old Raven Rock Covered Bridge on August 25, 1933, as Delaware floodwaters reached the highest level since the great flood of 1903, which had destroyed a span. The Delaware Canal should be visible at lower right, but the rising waters have covered the usual land separation between river and canal.

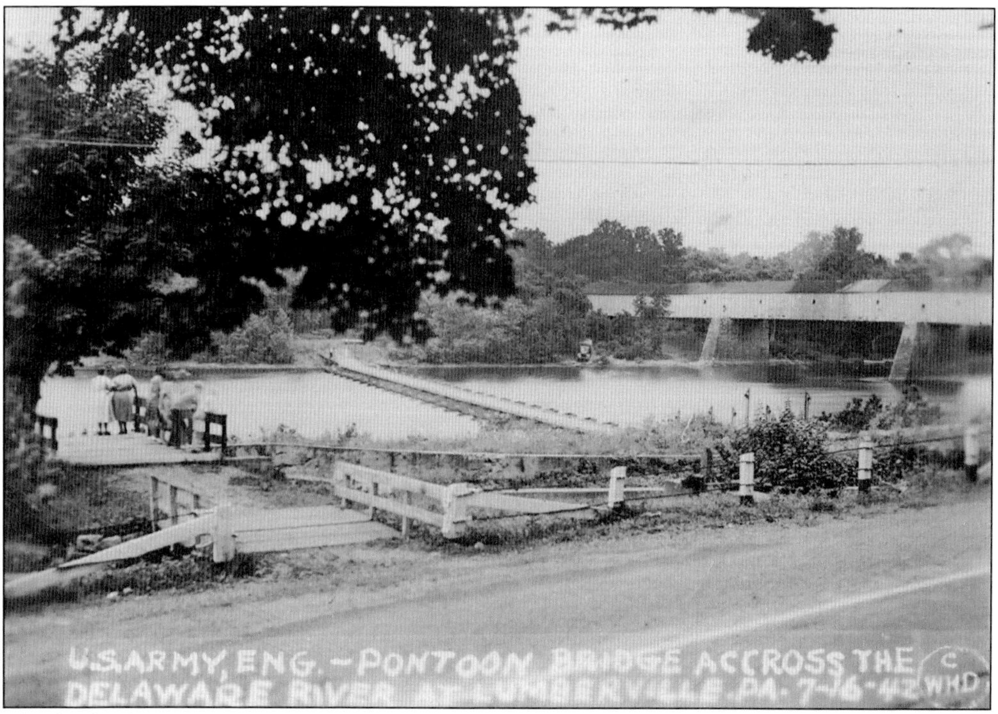

The bridge lost one span in the flood of 1903, replaced with steel. In 1932, the state agency purchased the bridge. Not trusting load limits, the military built a pontoon bridge next to the covered bridge for use during World War II. In 1947, a Roebling suspension bridge was built strictly for pedestrian passage.

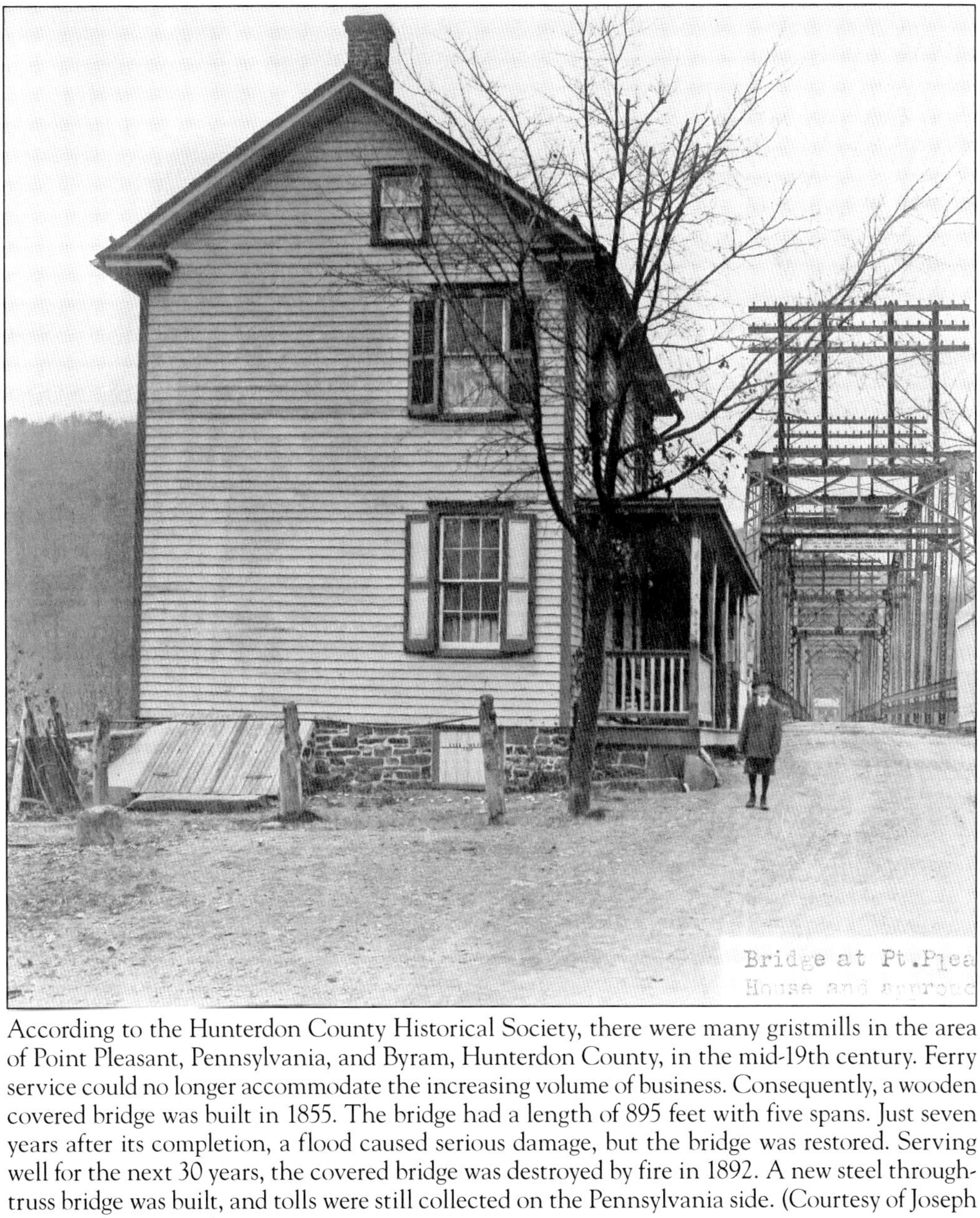

According to the Hunterdon County Historical Society, there were many gristmills in the area of Point Pleasant, Pennsylvania, and Byram, Hunterdon County, in the mid-19th century. Ferry service could no longer accommodate the increasing volume of business. Consequently, a wooden covered bridge was built in 1855. The bridge had a length of 895 feet with five spans. Just seven years after its completion, a flood caused serious damage, but the bridge was restored. Serving well for the next 30 years, the covered bridge was destroyed by fire in 1892. A new steel through-truss bridge was built, and tolls were still collected on the Pennsylvania side. (Courtesy of Joseph Donnelly, DRJTBC.)

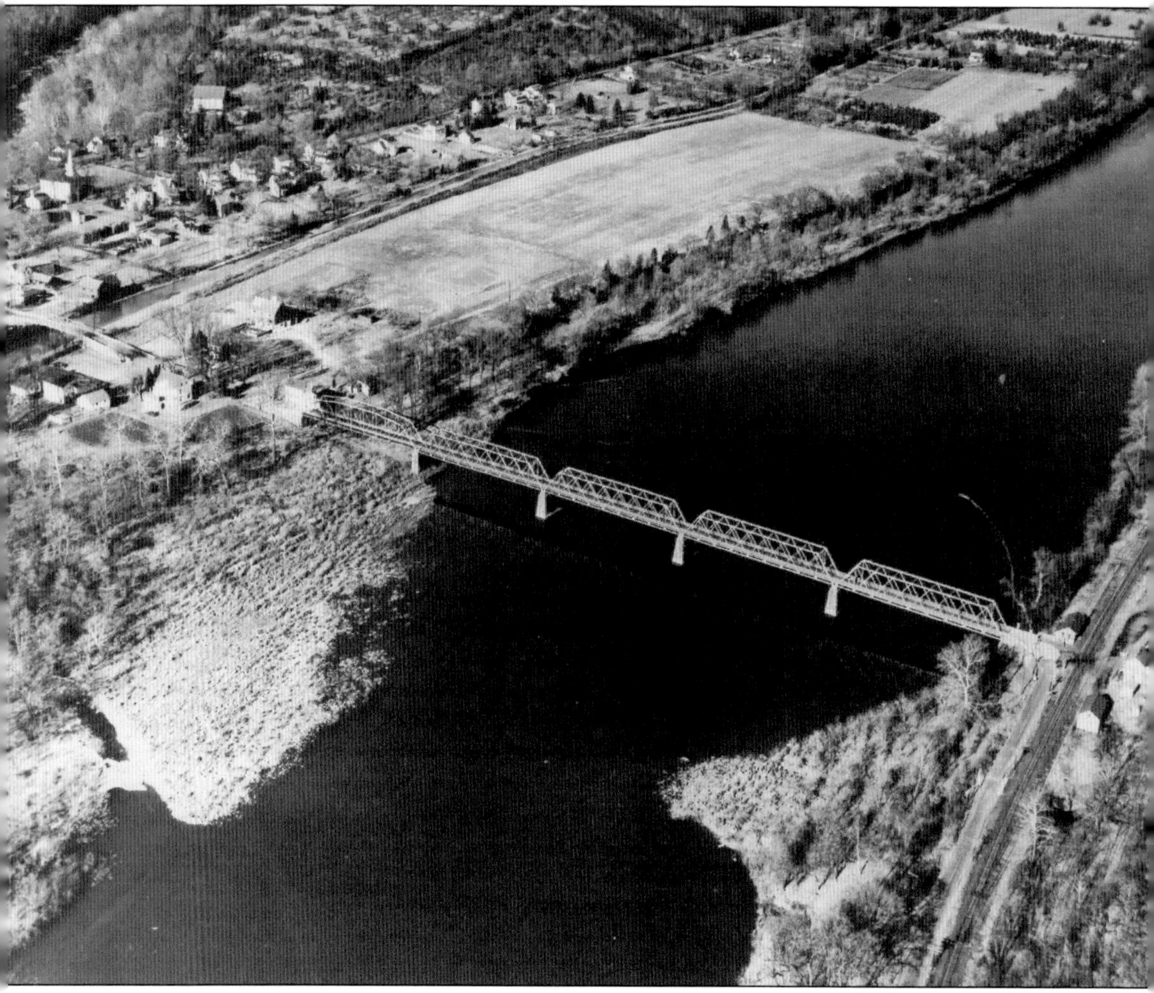

The Point Pleasant Delaware Bridge Company constructed the original covered bridge between Point Pleasant, Pennsylvania, and Byram, New Jersey. It provided Pennsylvania farmers with access to the new railroad built in New Jersey. In 1919, the Joint Bridge Commission purchased the bridge, eliminated all toils, and continued to maintain the five-span bridge. This bridge was destroyed by floodwaters from Hurricane Diane in 1955 and only the stone piers remain as testimony to this bridge. (Courtesy of Joseph Donnelly, DRJTBC.)

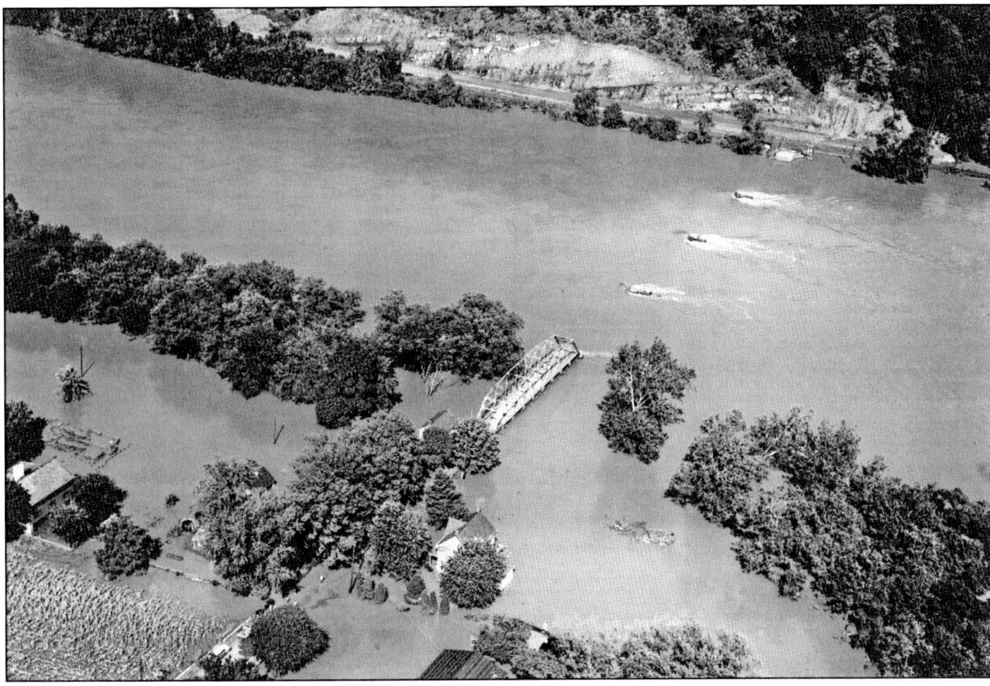

An aerial photograph taken shortly after the flood on the Delaware River caused by Hurricane Diane in 1955 shows the destruction of the Point Pleasant–Byram Bridge. In the *Bucks County Herald*, Mary Shafer wrote that summer colonies from Belvidere to Byram took massive hits from the rampaging river, some losing 95 percent of their dwellings. (Courtesy of the New Jersey State Archives, Department of State.)

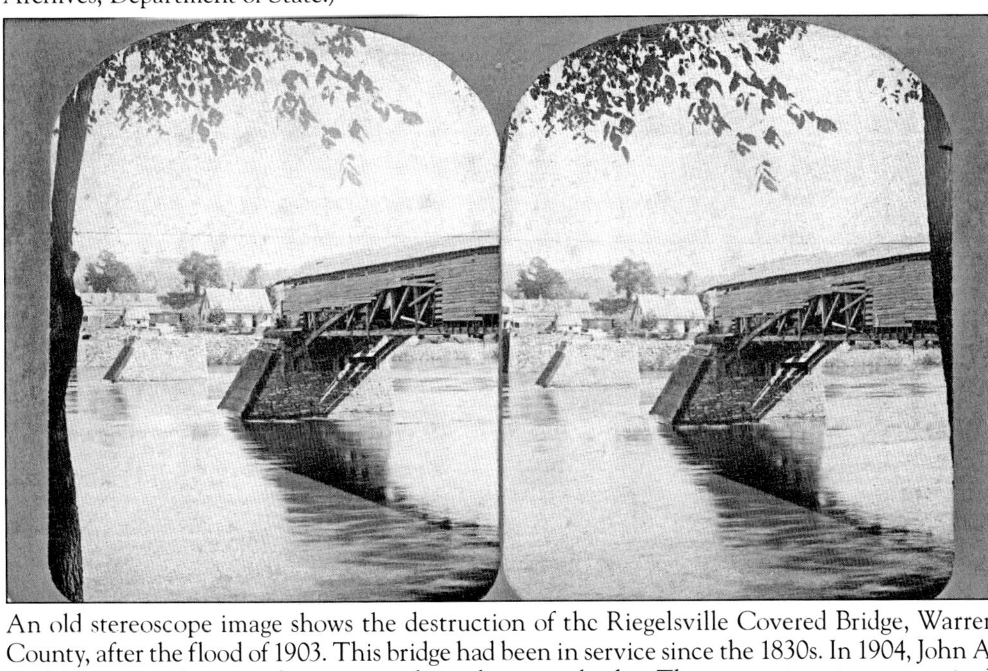

An old stereoscope image shows the destruction of the Riegelsville Covered Bridge, Warren County, after the flood of 1903. This bridge had been in service since the 1830s. In 1904, John A. Roebling Sons of New York constructed a replacement bridge. The supporting piers were raised, and a steel suspension bridge was erected. (Courtesy of Todd Clark.)

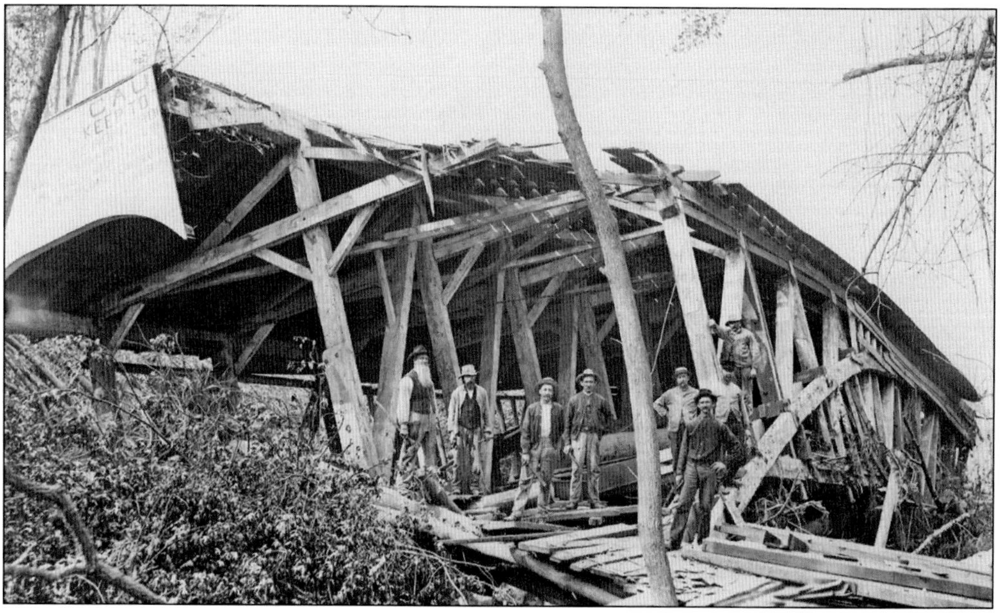

Benjamin Riegel was a founder of the Riegelsville Delaware Bridge Company, which opened a three-span covered bridge at the site of the former Shenk's Ferry in 1835–1836. It was a Burr-arch, multiple-kingpost structure approximately 577 feet long. The bridge carried horses, wagons, and pedestrians and served as an important link for family business interests on both sides of the river. The bridge was damaged in 1841, but it was repaired and reopened. The entire superstructure was swept away by the flood of 1903. These two photographs show the remains of the Riegelsville Bridge, relocated "10 miles from home," beached on Eckels Island with some of the men preparing to dismantle the ruined bridge for future use elsewhere. (Both, courtesy of the Hunterdon County Historical Society.)

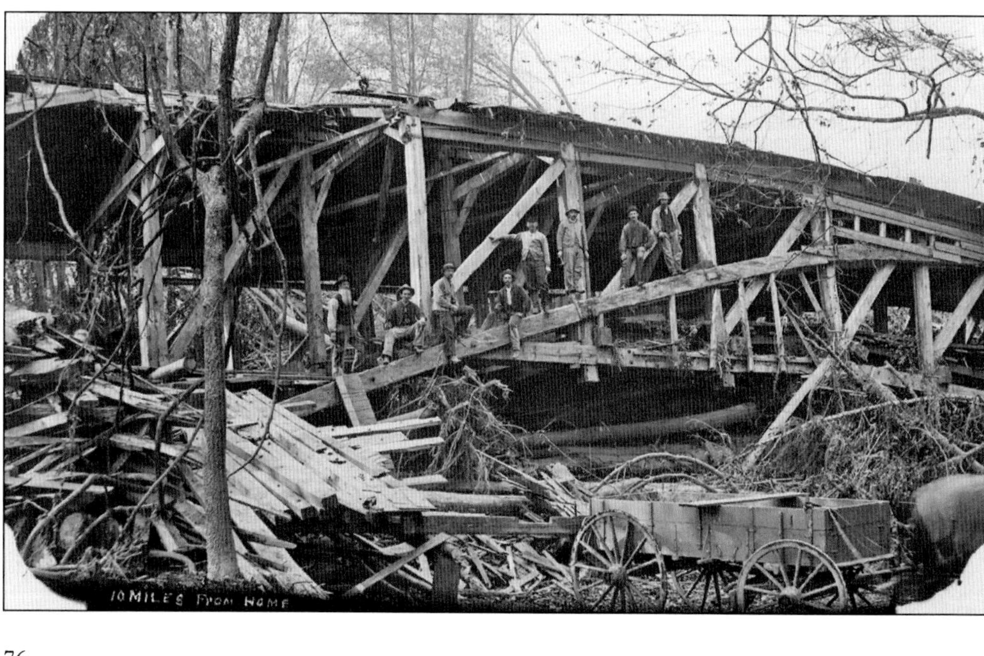

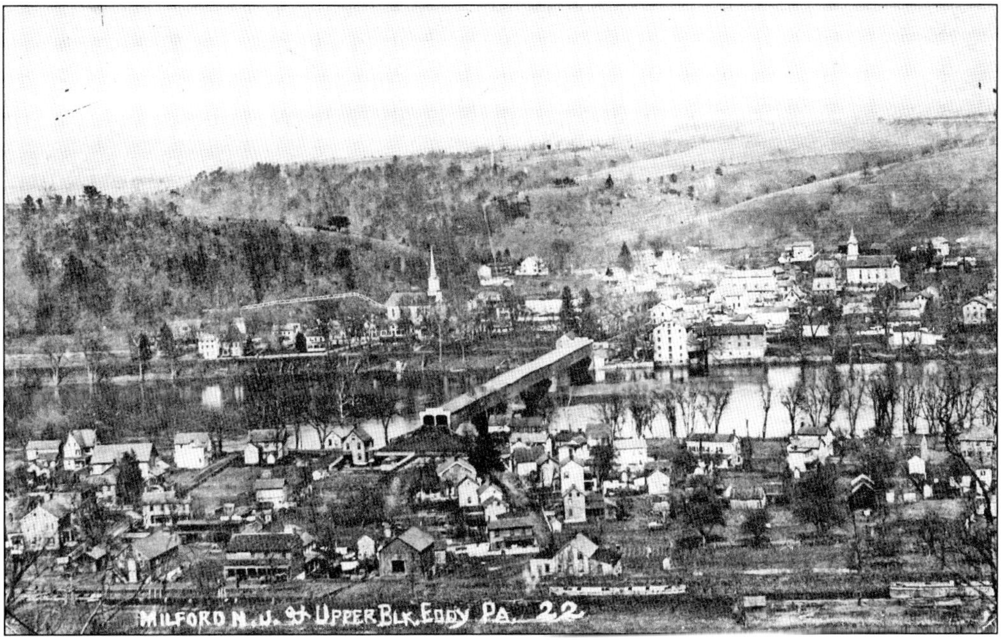

Milford, Hunterdon County, was originally settled in the 1750s as part of a proprietary tract owned by Col. John Reid of New York, who never actually lived there. By 1760, Milford served as an active ferry crossing. The increase in population and the building of a second sawmill spurred the construction of the covered bridge and the consequent need for a toll collector.

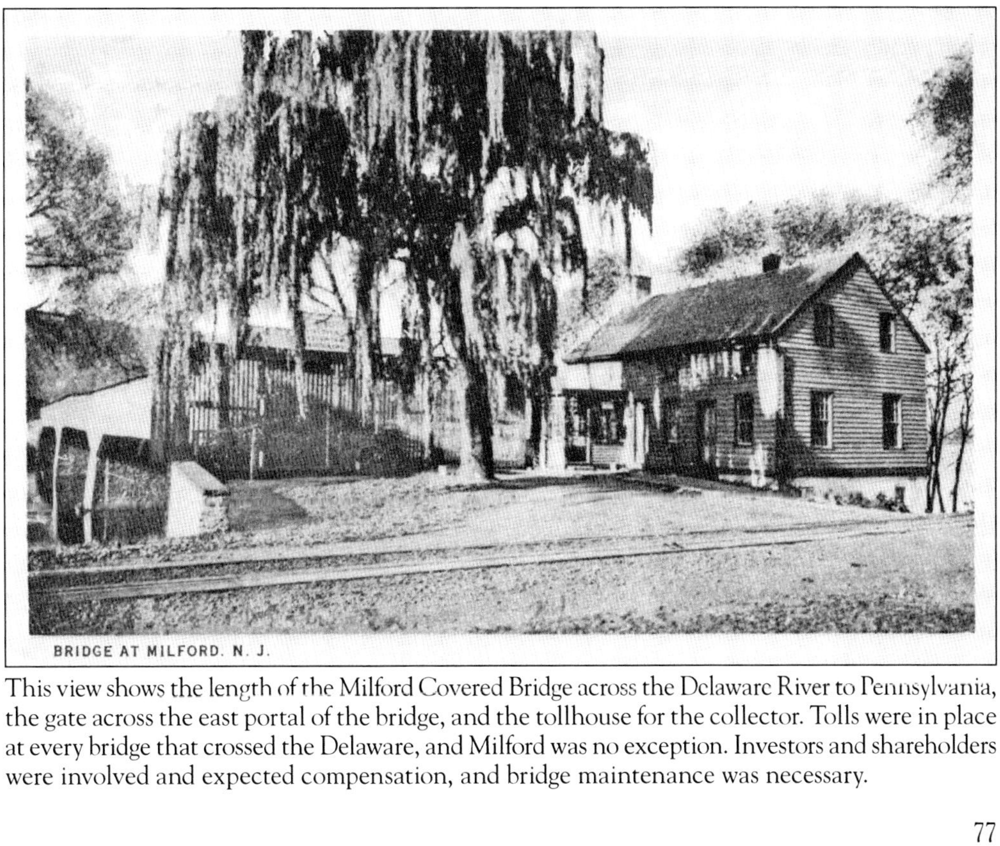

This view shows the length of the Milford Covered Bridge across the Delaware River to Pennsylvania, the gate across the east portal of the bridge, and the tollhouse for the collector. Tolls were in place at every bridge that crossed the Delaware, and Milford was no exception. Investors and shareholders were involved and expected compensation, and bridge maintenance was necessary.

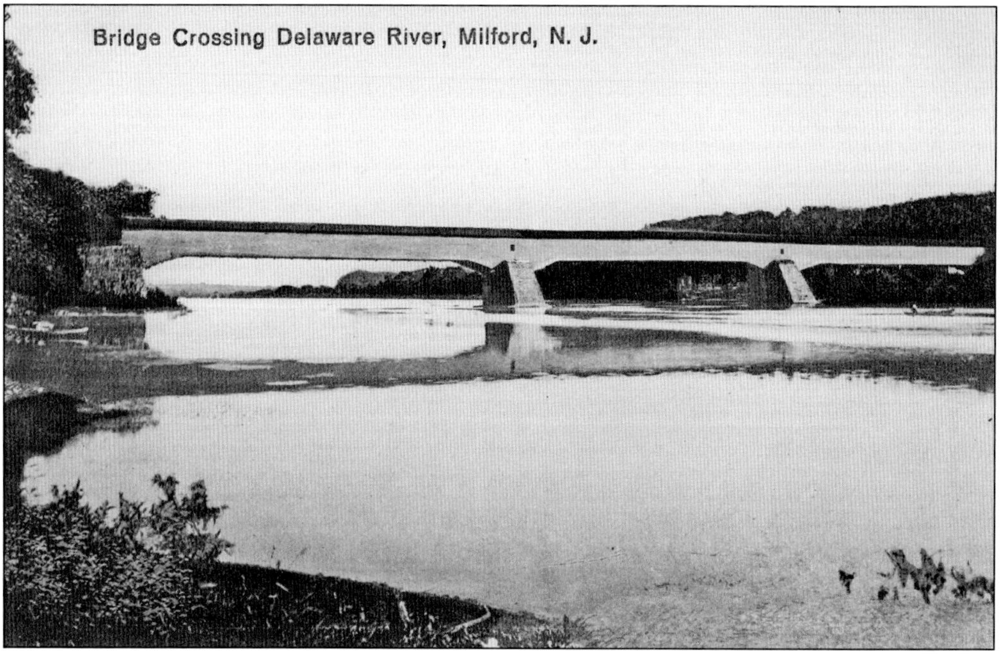

A bridge was usually built across a river when demand exceeded what a ferry could provide. At Milford, a bridge was proposed to cross the Delaware River. The Milford Delaware Bridge Company received approval in 1839. The Milford Covered Bridge was only 681 feet long, shore to shore. It consisted of three spans supported by two piers. This made it safer for timber rafts to pass underneath in the rushing waters of a spring freshet. The arch and truss combination of Theodore Burr was used. On January 29, 1842, the bridge was completed and opened as a toll bridge. The initial cost of building the bridge was only $27,000. Below, the toll keeper's house is shown.

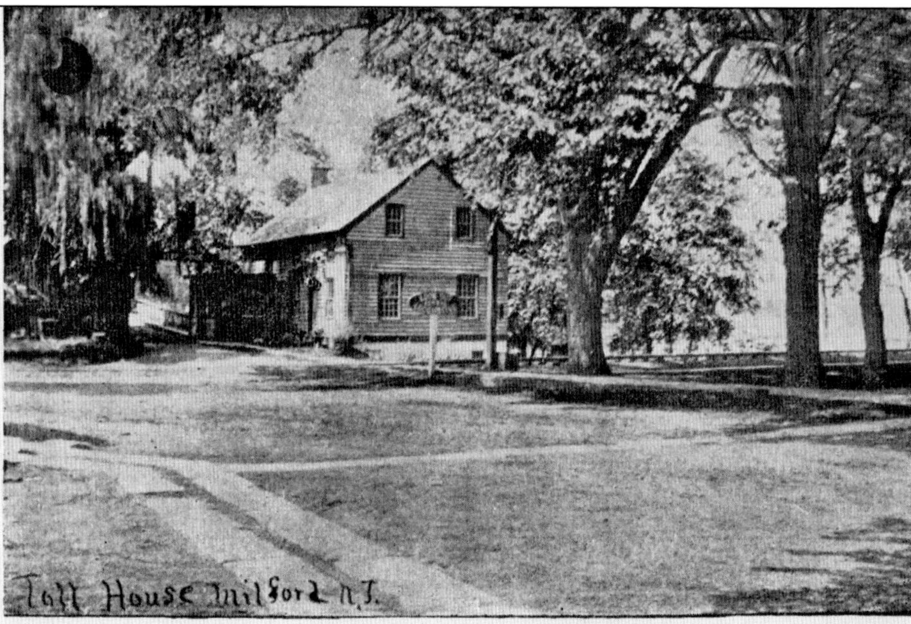

Toll House, Milford, N. J.

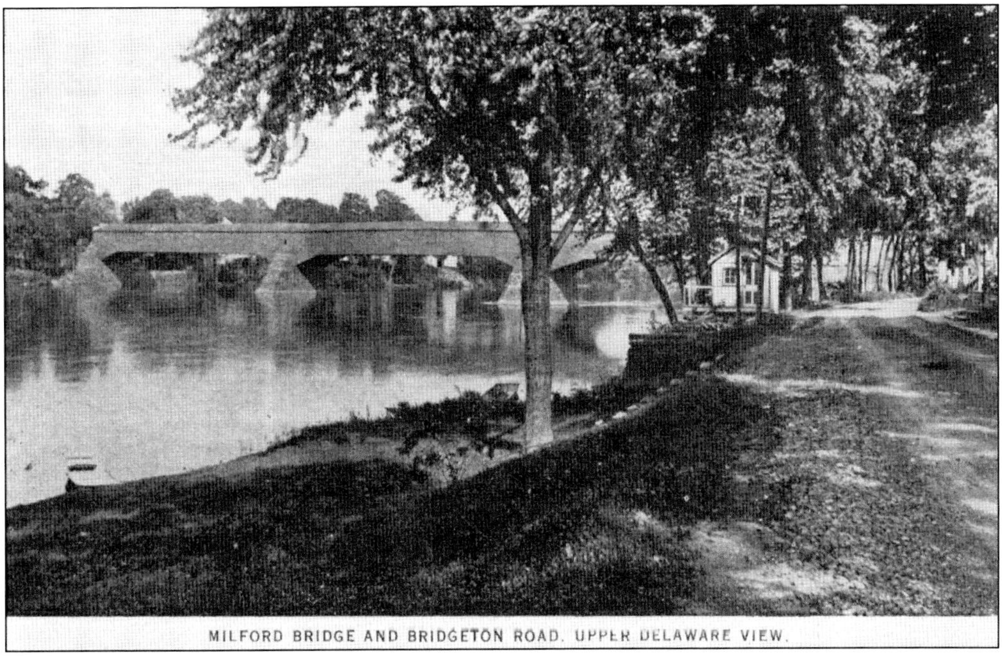

This view of the bridge comes from upriver in Bridgeton Township on the Pennsylvania side, above the town of Upper Black Eddy, or UBE, as locals refer to it. The arches from the Burr-truss design can be seen embedded into the piers and abutment on the New Jersey side. All three spans are intact.

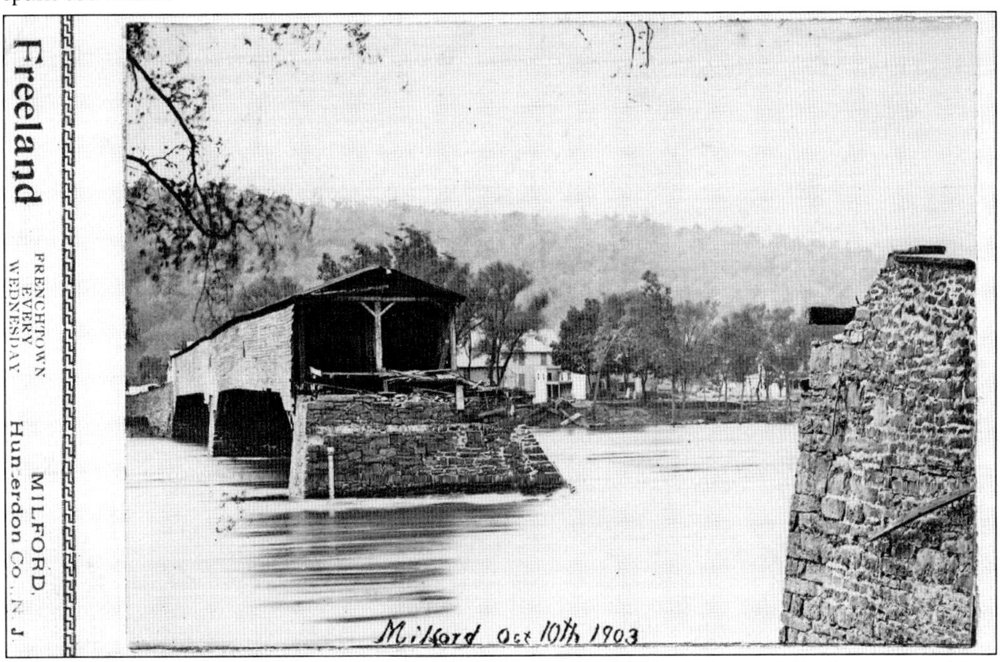

Following the flood of October 10, 1903, the span of the Milford Covered Bridge nearest New Jersey was destroyed and swept away. It was quickly rebuilt using the same design and with timbers salvaged from the destroyed Riegelsville Covered Bridge upriver. On June 28, 1929, the Joint Bridge Commission purchased the bridge and eliminated all tolls.

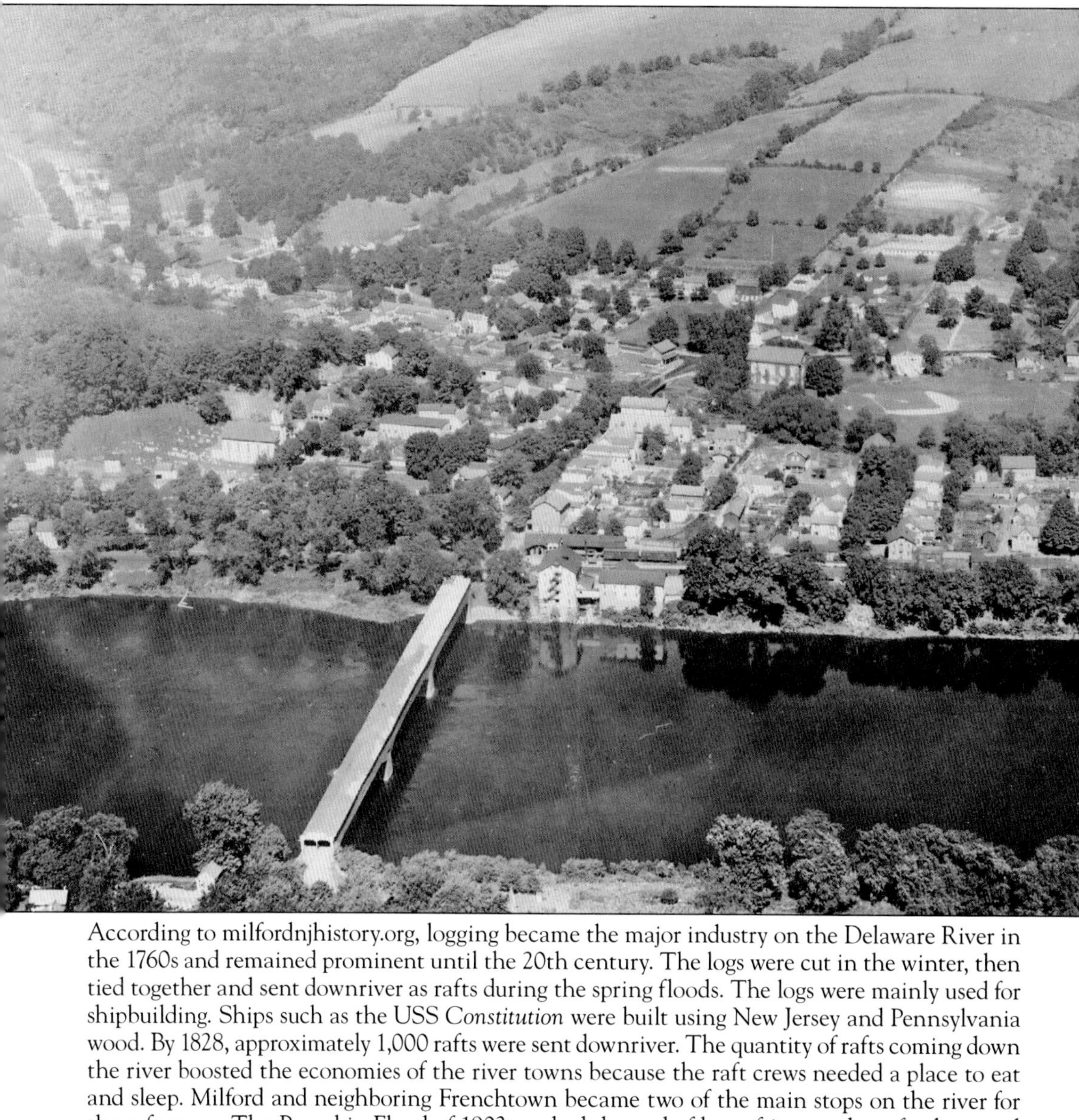

According to milfordnjhistory.org, logging became the major industry on the Delaware River in the 1760s and remained prominent until the 20th century. The logs were cut in the winter, then tied together and sent downriver as rafts during the spring floods. The logs were mainly used for shipbuilding. Ships such as the USS *Constitution* were built using New Jersey and Pennsylvania wood. By 1828, approximately 1,000 rafts were sent downriver. The quantity of rafts coming down the river boosted the economies of the river towns because the raft crews needed a place to eat and sleep. Milford and neighboring Frenchtown became two of the main stops on the river for the raft crews. The Pumpkin Flood of 1903 marked the end of log rafting, as the rafts destroyed several bridges, including the Milford Bridge.

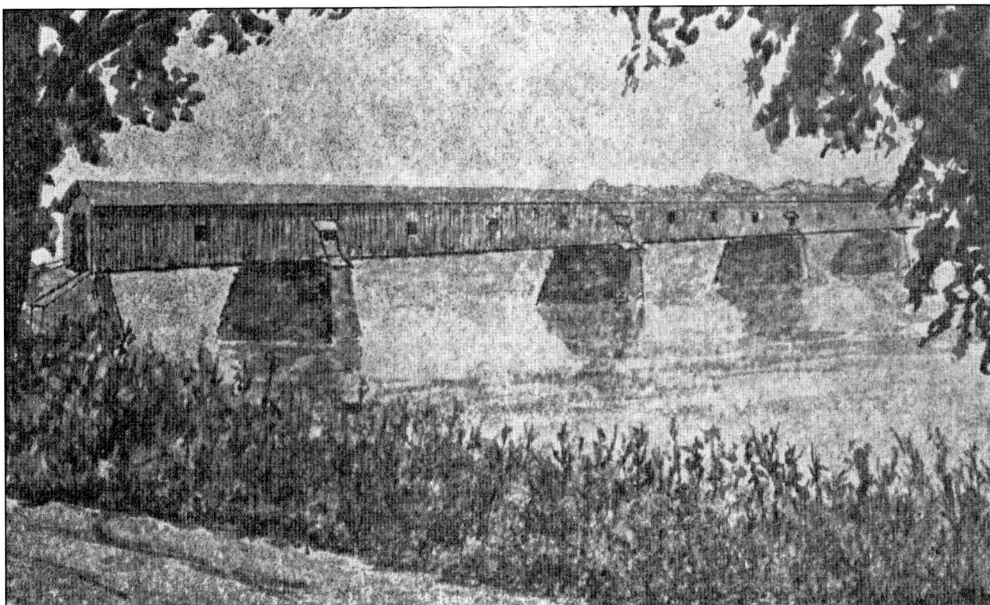

Former Covered Bridge, Frenchtown, N. J. — By Ethel D. Hoffman

The Frenchtown-Uhlerstown Covered Bridge carried Bridge Street traffic from Frenchtown, Hunterdon County, to Uhlerstown, Tinicum Township, Pennsylvania. The original bridge was constructed after 1841. It consisted of six covered wooden spans using the latticed Town-type trusses. The masonry substructures of the bridge—five piers and two abutments—are still standing and now support the present bridge. The painting above is by Ethel D. Hoffman, a 20th-century American artist. Frenchtown grew substantially in the first half of the 19th century. A trade in grain and other farm products developed, and many mills were located on or near the river. Later, mills were converted into larger factories, and boatmen and river men also helped fuel Frenchtown's growth. (Both, courtesy of the New Jersey Historical Society.)

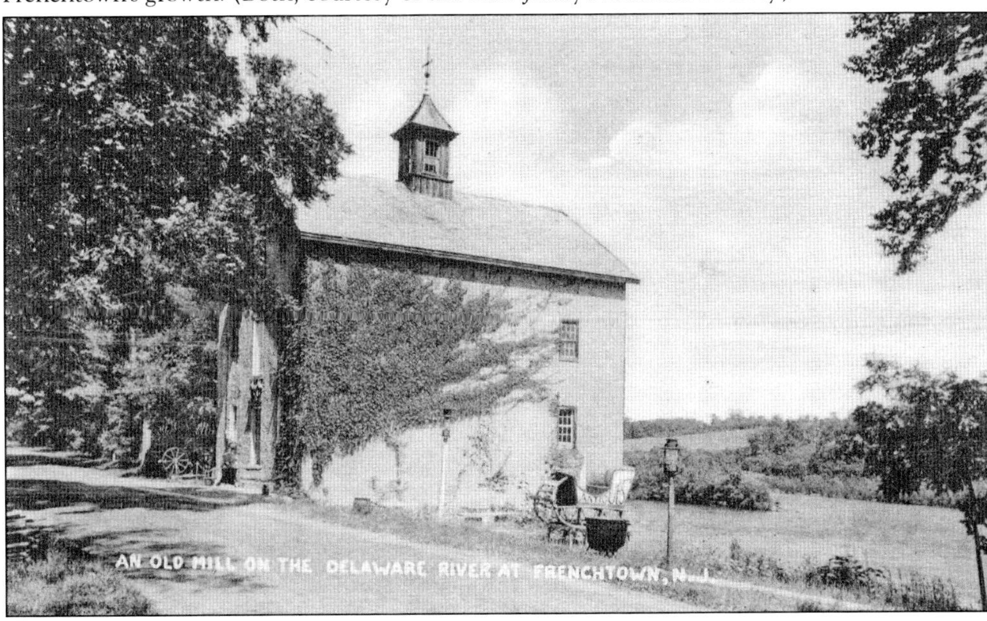

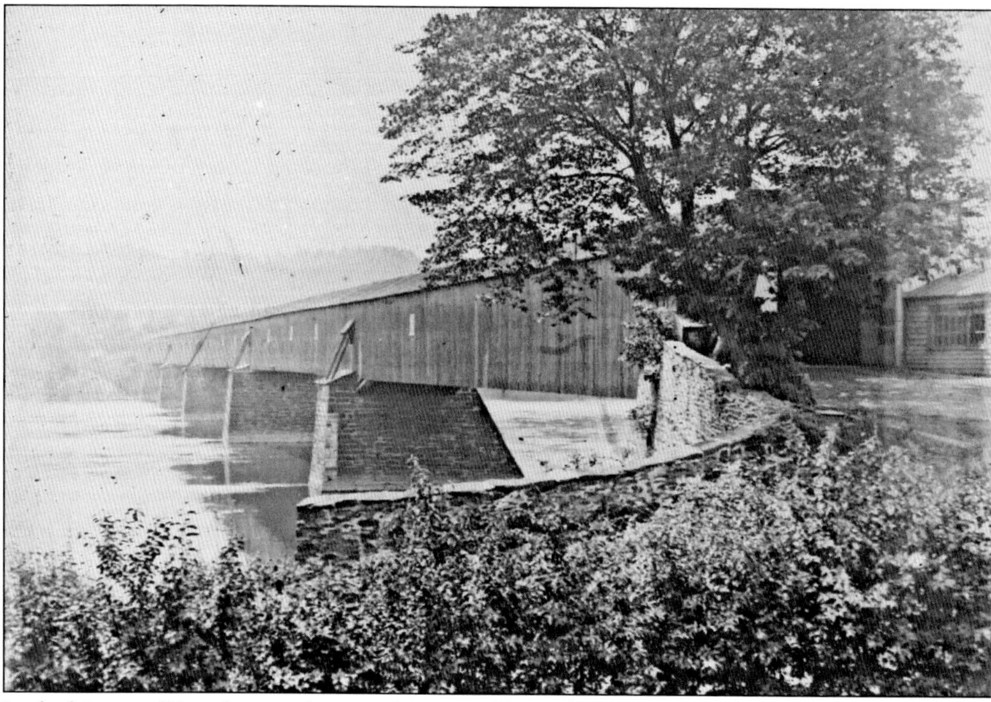

In the history of Frenchtown, the years between 1794 and 1836 are often referred to as the Malletian Era. Paul Henri Mallet purchased all the land that is now Frenchtown from Thomas Lowrey and built some of the first homes. His sons continued to sell parcels of land throughout the town, which later became known as Frenchtown because of Mallet, who was erroneously thought to be French. The above picture shows the entire bridge structure, 962 feet long, which did survive an 1862 flood. During the Pumpkin Flood of 1903, however, two spans on the New Jersey side were lost, and the owners replaced them with two steel-truss spans, as seen in the picture below. (Above, courtesy of Todd Clark; below, author's collection.)

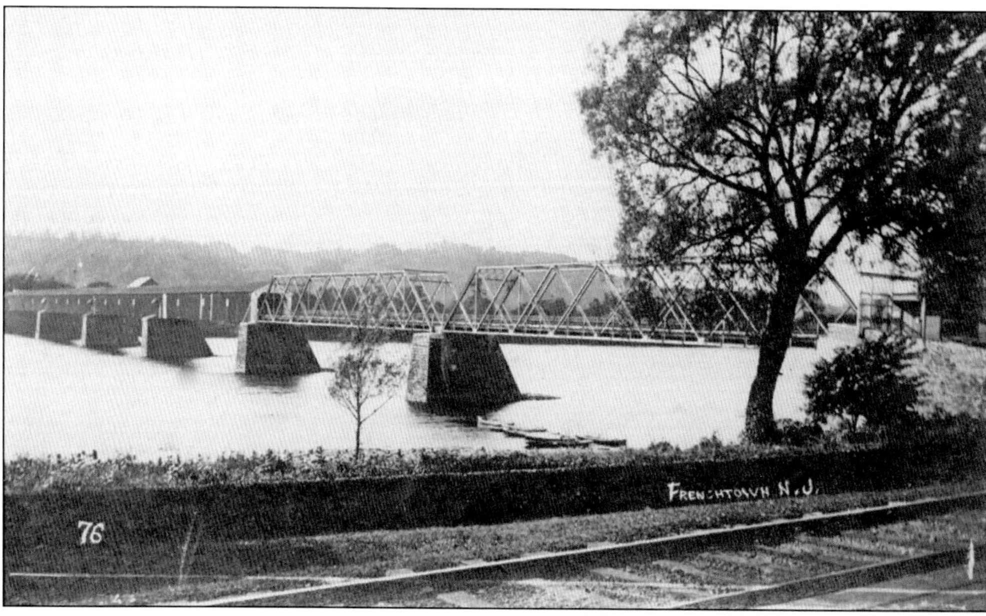

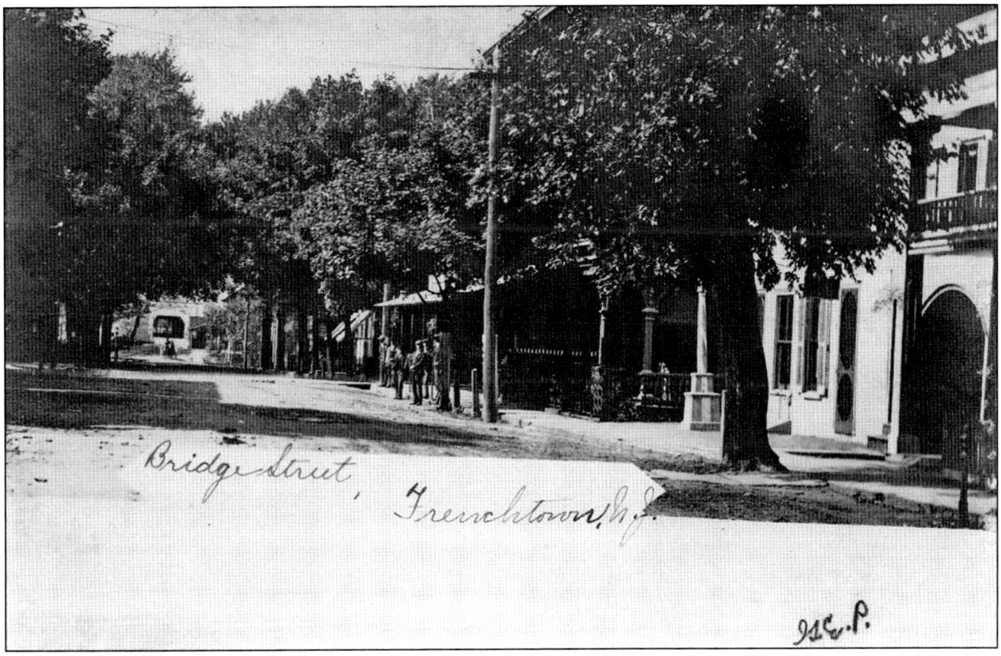

The first bridge across the Delaware River at Frenchtown was a six-span wooden covered bridge, begun in 1843 on the five piers that still stand today. Thus, the town became a gateway to Pennsylvania. The bridge was just near 1,000 feet in length and opened in 1844, avoiding the devastating flood of 1841. It survived the flood of 1862 and a terrible fire that consumed most of Bridge Street. Losing two spans on the New Jersey side in the major flood of 1903, the bridge was rebuilt with two steel trusses and remained as such until the Delaware River Joint Bridge Commission purchased it in 1929. The commission then replaced it with a steel six-span Warren-truss bridge in 1931.

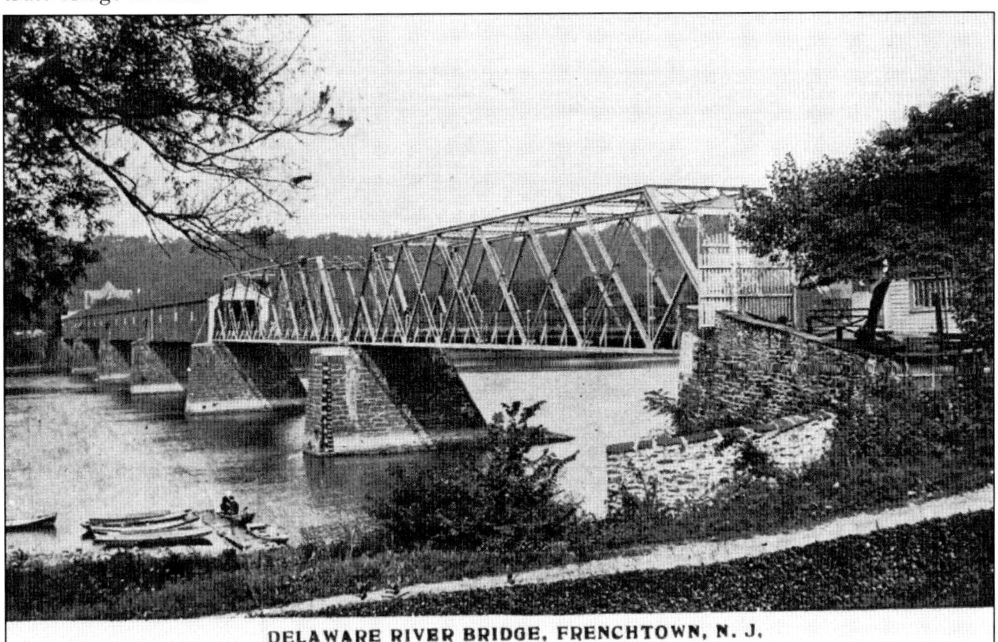

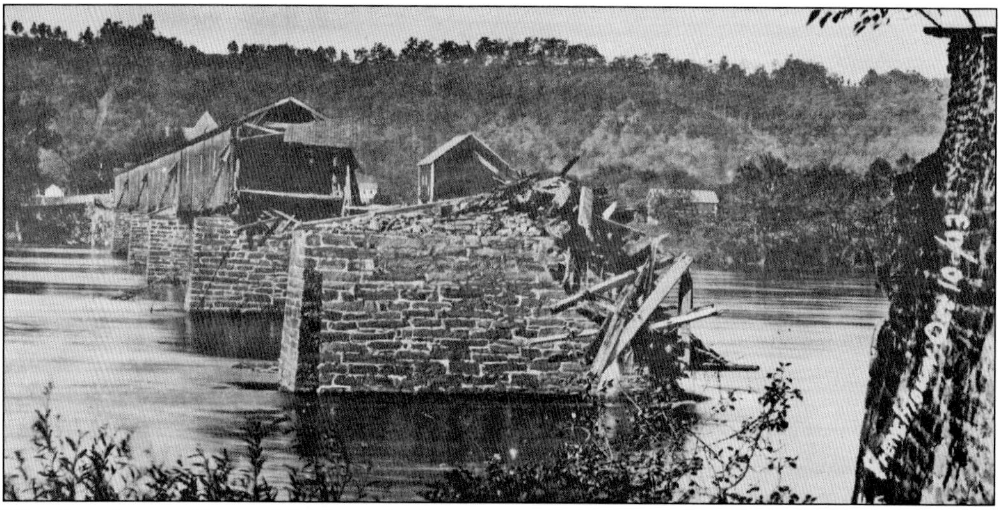

The Frenchtown Covered Bridge, which connected to Uhlerstown, Pennsylvania, opened in 1844. It survived quite well during an 1862 flood and a fire on Bridge Street in 1878. The covered bridge was less fortunate on October 10, 1903, when another major flood raced downriver and destroyed the two spans nearest the New Jersey riverside.

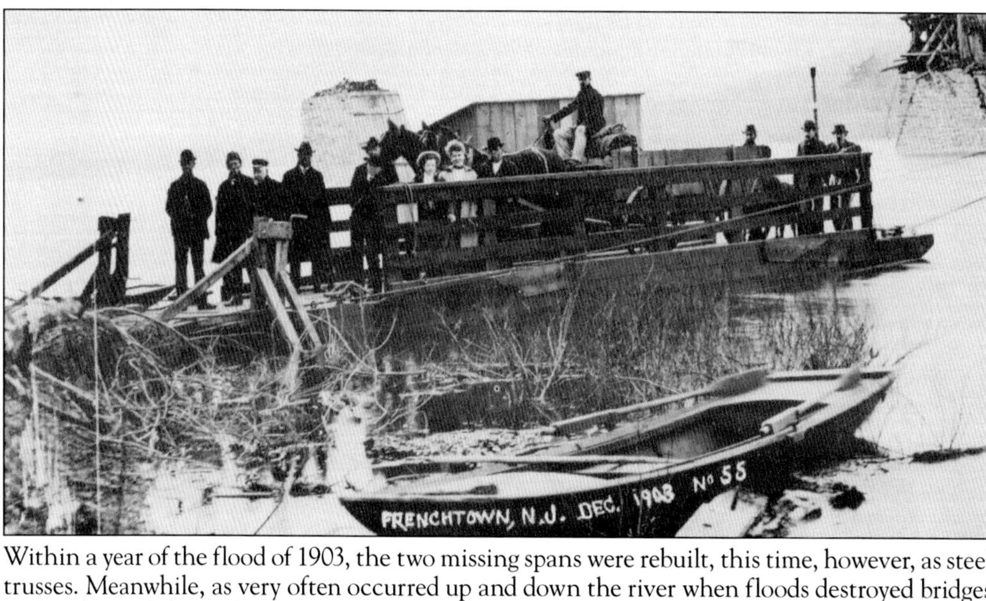

Within a year of the flood of 1903, the two missing spans were rebuilt, this time, however, as steel trusses. Meanwhile, as very often occurred up and down the river when floods destroyed bridges or fires burned them, the ferry came back into service. Many ferry operators lived on both sides of the river, ready to start up service.

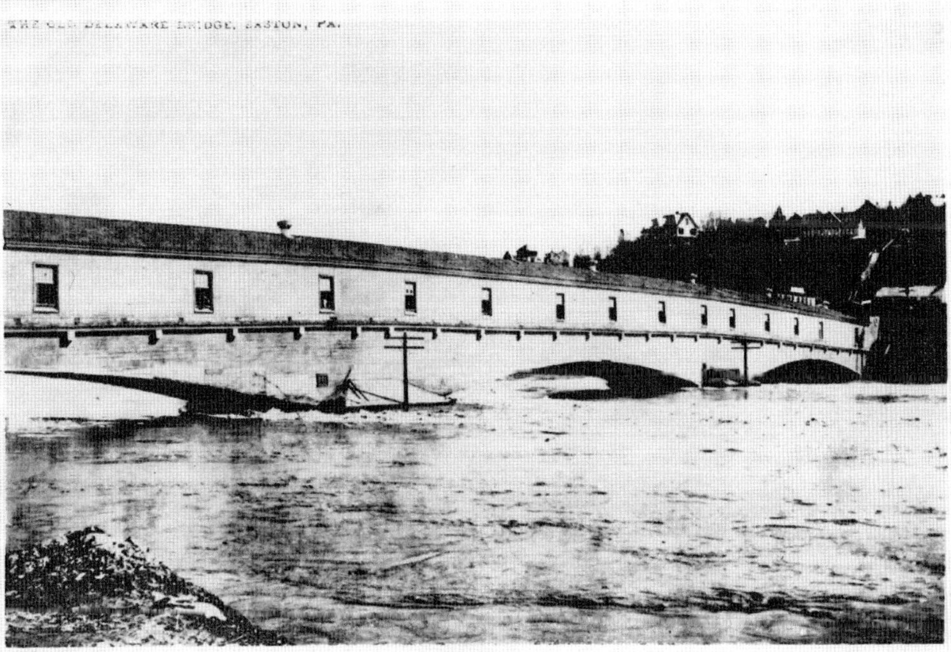

The first crossing of the Delaware River at Phillipsburg, Warren County, and Easton, Pennsylvania, occurred in 1739, with a ferry operated by David Martin. This ferry service continued until the beginning of the 19th century but was eventually replaced with a wooden covered bridge that was opened to traffic in 1806. The bridge was designed and built by Timothy Palmer, one of the foremost bridge builders of his time. Called the Northampton Street Bridge, it was 550 feet long over three spans and had a 29-foot-wide roadway through trussed arches below. Tolls were reduced in 1845, and pedestrian tolls were removed in 1856. The covered bridge was replaced in 1895. In the above photograph, note the level of the river compared to that below.

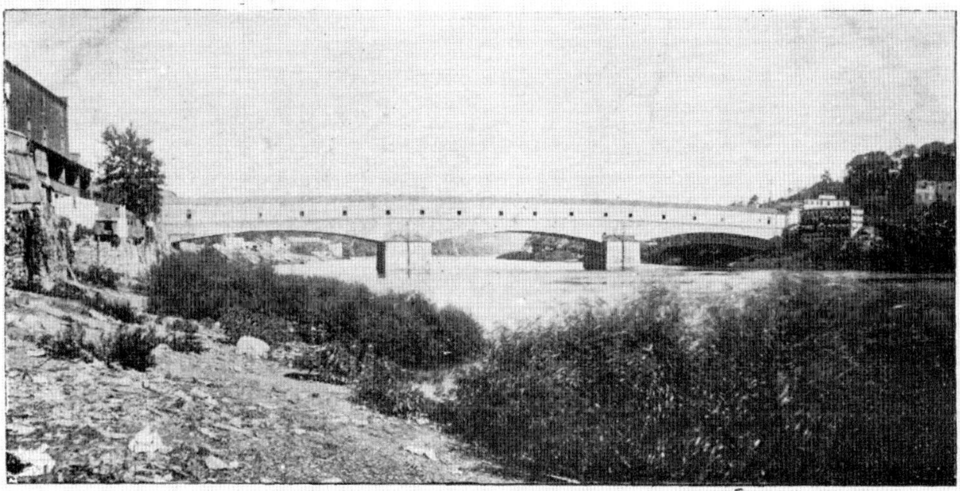

The Old Easton Bridge. Erected 1805—View from the "Forks."

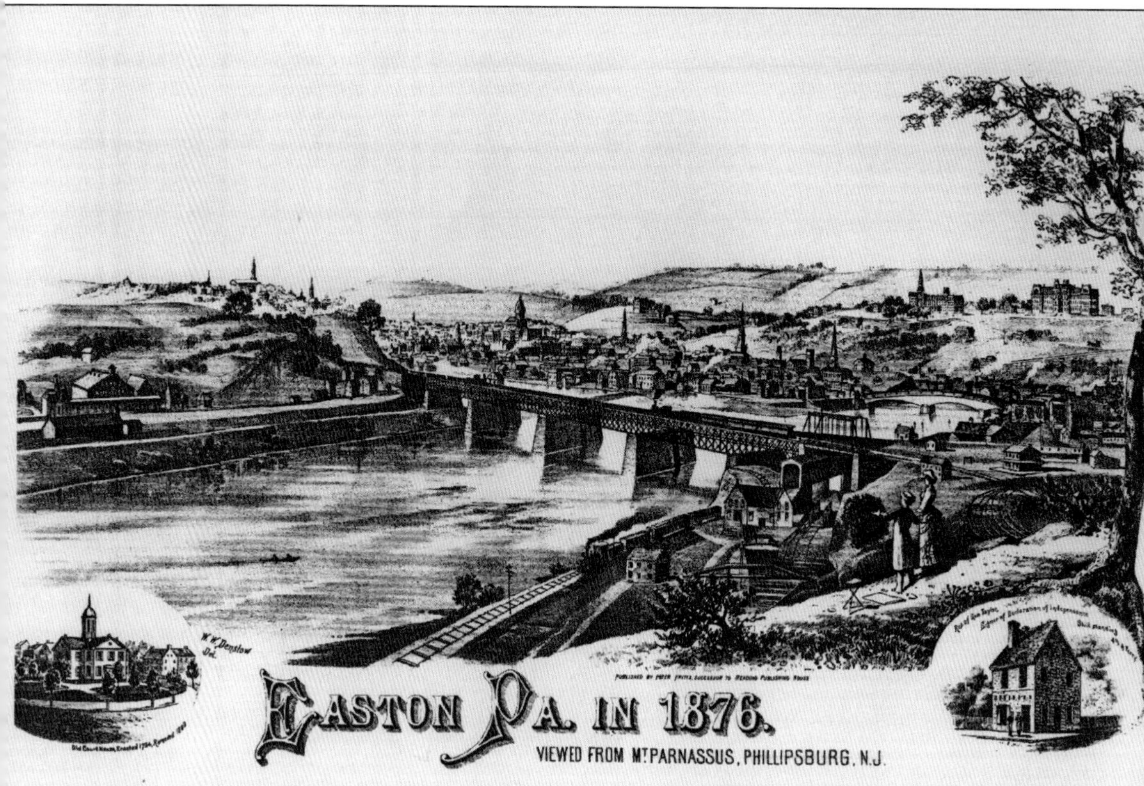

Both the Delaware and Lehigh Rivers hosted great volumes of commerce, given the junction of the Morris, Lehigh, and Delaware Canals. Coal was transferred from railcars to boats at Port Delaware to supply the Morris Canal. In the past, routes of the Lehigh Valley, Jersey Central, Pennsylvania, Lackawanna & Western and Lehigh & Hudson River Railroads met and interchanged freight and passenger traffic here. At one time, there were five major railroad yards, eight roundhouses on turntables, and five passenger and freight stations in the vicinity. Ingersoll-Rand in Phillipsburg built the first diesel-electric locomotive. The double-decked bridge was built of wood in 1856 and was the first railroad bridge over the Delaware River at Phillipsburg. This bridge connected three railroads, including two that used the bridge, the Central Railroad of New Jersey and the Lehigh Valley Railroad. Palmer's Northampton Street Covered Bridge can be seen upriver, in the center right of this photograph, crossing the Delaware River. Near the double-decked bridge, parts of covered railroad bridges can be seen near the canal and river.

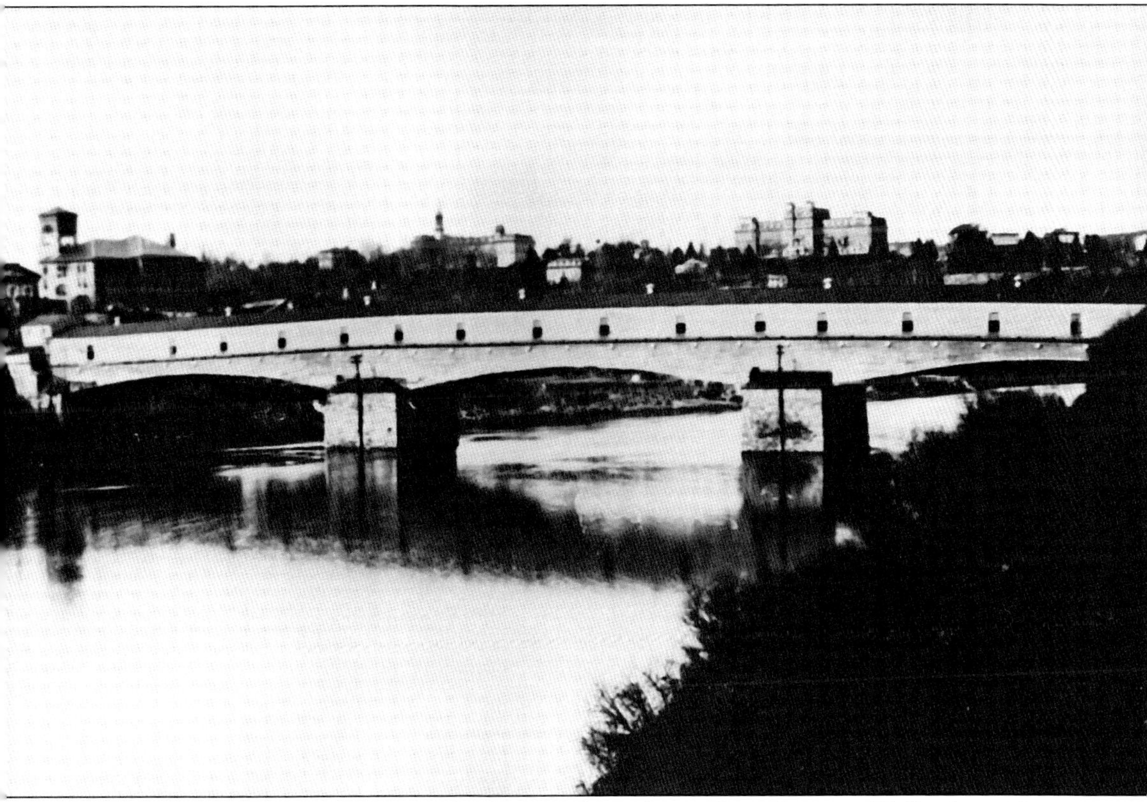

Timothy Palmer's three-span covered bridge over the Delaware River between Phillipsburg, New Jersey, and Easton, Pennsylvania, was very similar in appearance to drawings of his Permanent Bridge built in Philadelphia, considered the first covered bridge in North America. In Phillipsburg, the predecessor of the modern trolley system was the Phillipsburg Horse Car Railroad Company, which was chartered in 1867 and built along South Main Street. In 1885, the line continued on to Easton. The Center Square and Delaware Passenger Railway, also a horse-drawn trolley, was chartered in 1871 and began service over the covered bridge in 1885. Enduring many floods and storms, the bridge finally weakened with the steel railway tracks installed for the streetcars and trolleys. A new bridge was built in 1895.

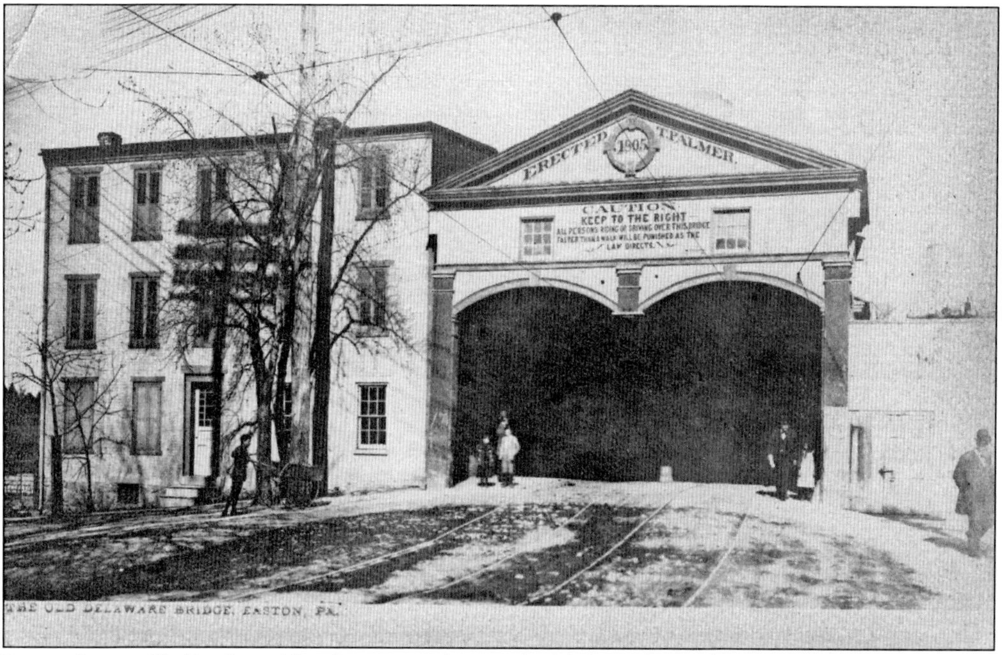

Proceeding to Phillipsburg, New Jersey, an elegant double-barrel portal greets the traveler entering the Easton-Phillipsburg, or Northampton Street, Covered Bridge. The writing on the peaked facade gives the date of construction, 1805, and the builder's name, T. Palmer. The toll keeper's house is seen on the left, and trolley tracks lead into the bridge. (Courtesy of Todd Clark.)

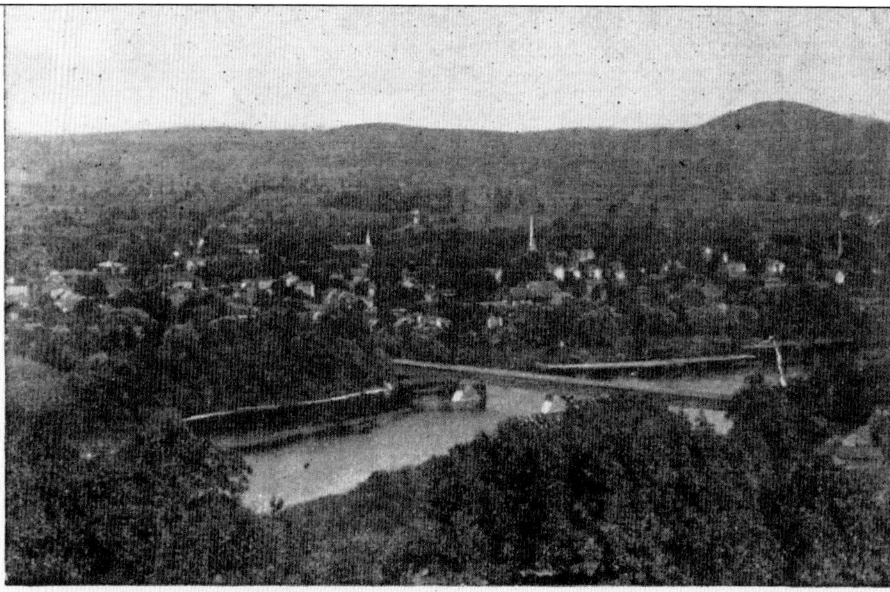

View of Belvidere, showing old covered bridge, destroyed by flood, Oct, 1903

Belvidere was part of Oxford Township, Warren County, until 1845, when it was organized as a borough. The business activity of Belvidere always depended in a great measure on the presence of the Delaware River, which offered possibilities of power and communication. The covered bridge served the area of Belvidere between the years of 1839 and 1903.

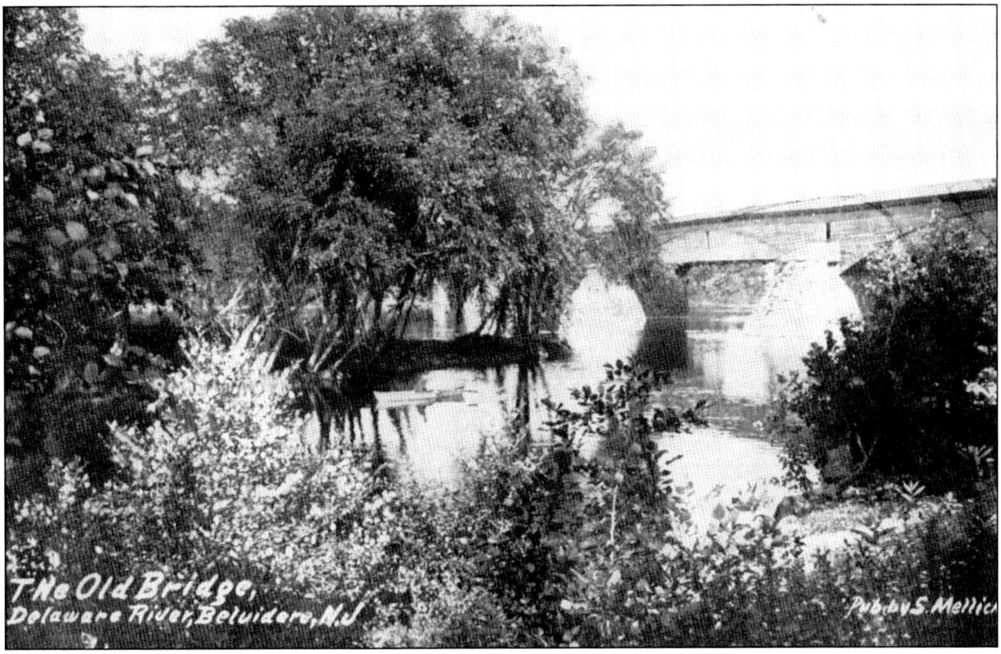

In the early 19th century, a local doctor on house calls charged his patients the ferry tolls. He became known as "Dr. Belvidere." This area of the river was considered dangerous. A bridge company was established by both states, and construction began in 1834. After its opening in 1836, the bridge was damaged by a storm, forcing it to close until 1839.

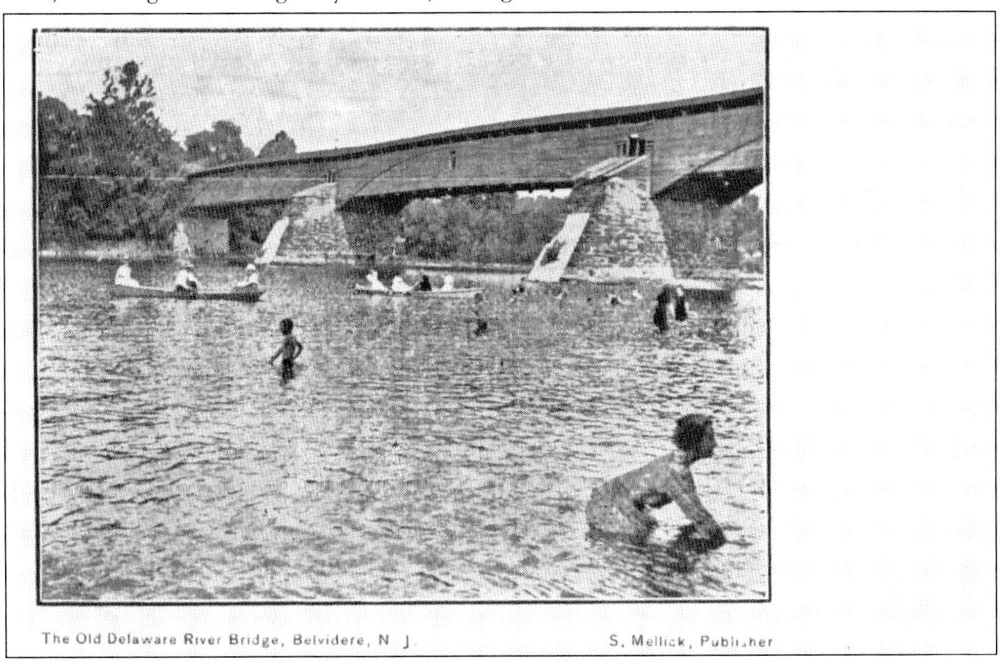

Swimming and boating were common activities in the waters of the Delaware River near the Belvidere Covered Bridge. Begun in 1834, the bridge itself was four spans built in the Burr-arch design by Solon Chapin. It was well over 600 feet long. Looking closely, the supporting arches can be seen alongside the bridge.

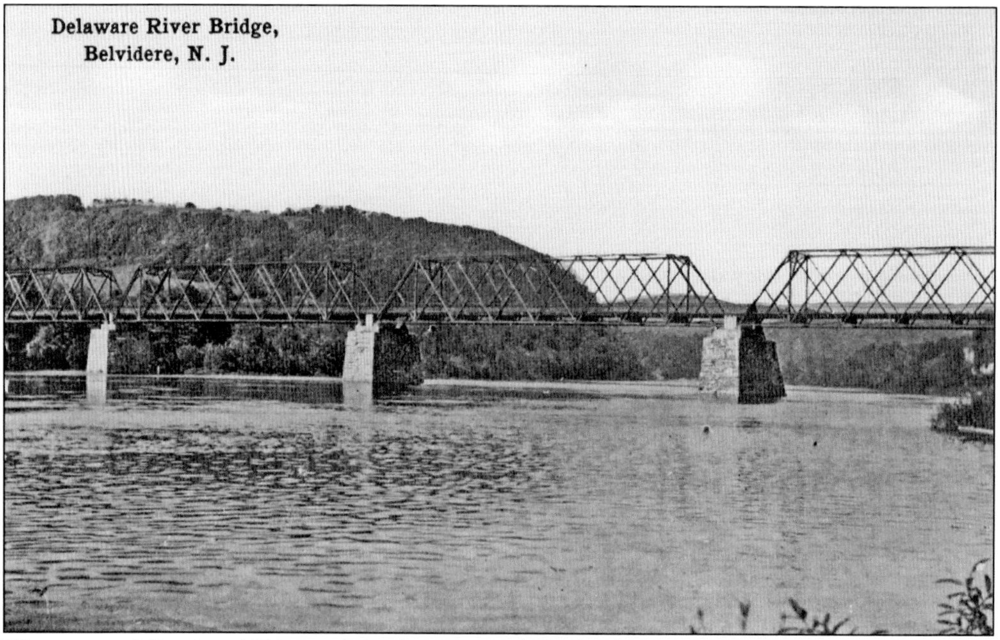

The storm of October 1903 struck both towns and the bridge in between. Many lives were lost. On that day, the toll collector closed off the gate and did not allow anyone to cross. At 6:40 that evening, a crash was heard, and seconds later, the bridge was no more. In 1904, a steel double Warren truss was constructed.

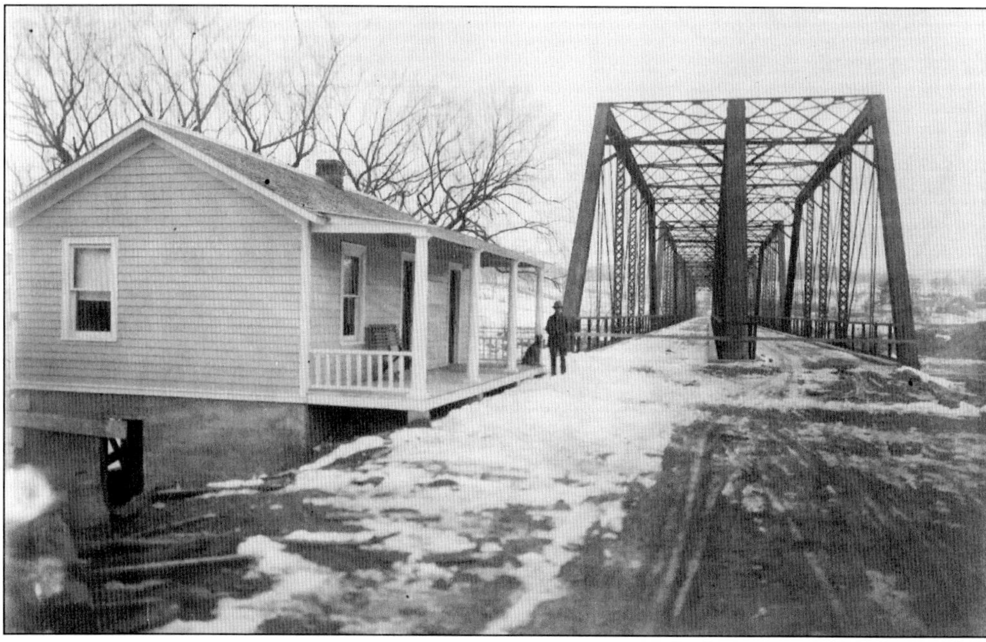

The Darlington Bridge at Delaware Station, Warren County, was a highway bridge over the Delaware River. It was formerly a metal railroad bridge constructed by the Delaware, Lackawanna & Western Railroad in 1871. The original bridge was a wooden covered bridge that had been built in 1855. The wooden covered bridge was constructed on the DL&W railroad mainline, with three spans and a single track. (Courtesy of Joseph Donnelly, DRJTBC.)

When the railroad built a new bridge next to its existing one, it put the old bridge up for sale. An Episcopal minister, Henry V.B. Darlington, purchased the bridge for $5,000, turned the structure into a highway toll bridge, bought out the local ferry, and carried on a very prosperous business for years. The 1954 demolition is shown. (Courtesy of Joseph Donnelly, DRJTBC.)

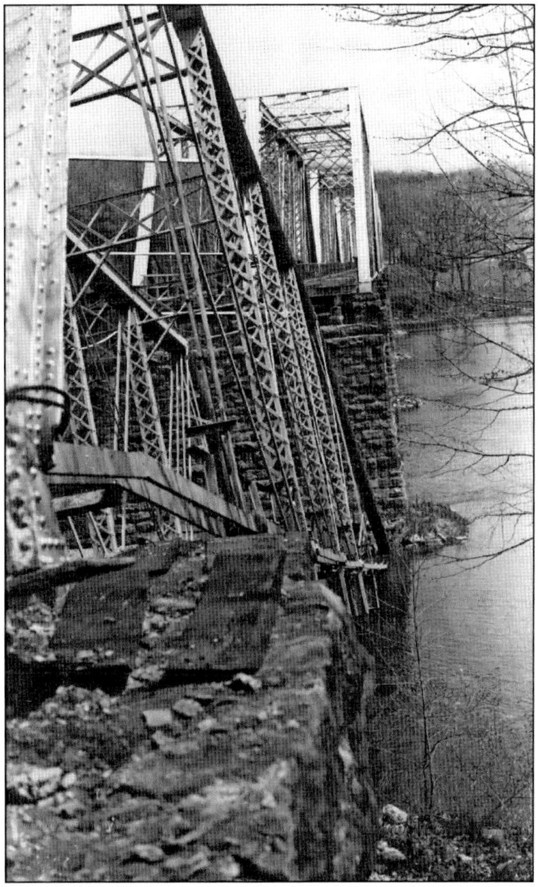

Charles Kellogg and Maurice Company constructed the original covered bridge structure between Columbia, Warren County, and Portland, Pennsylvania, as a vehicular bridge with an 18-foot roadway. The entire bridge was protected from the weather by wooden sides and a slate roof. Early on, travelers considered it the "Gateway to the Poconos."

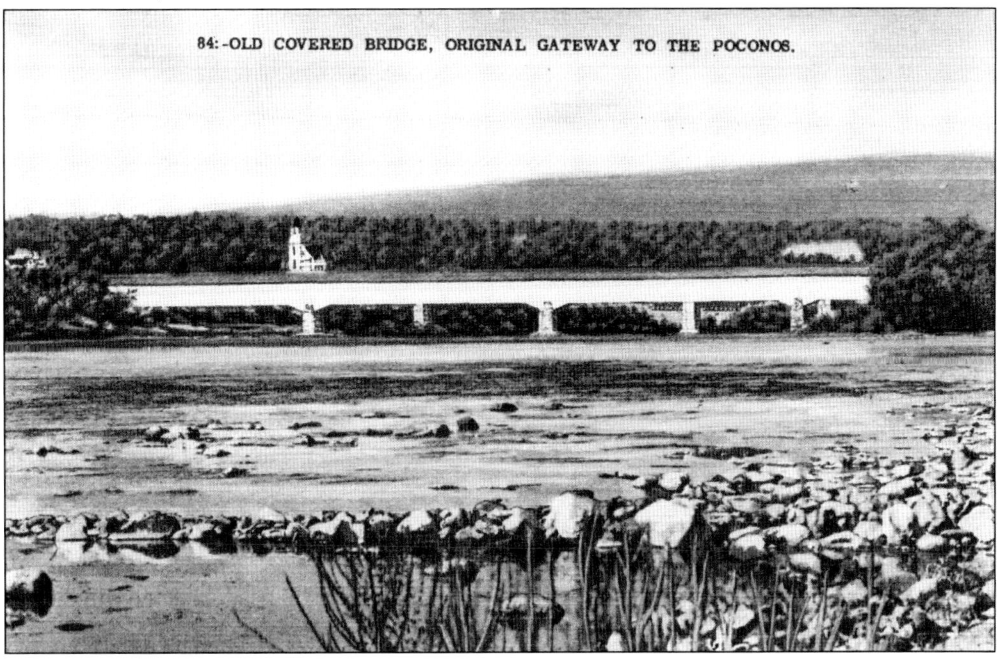

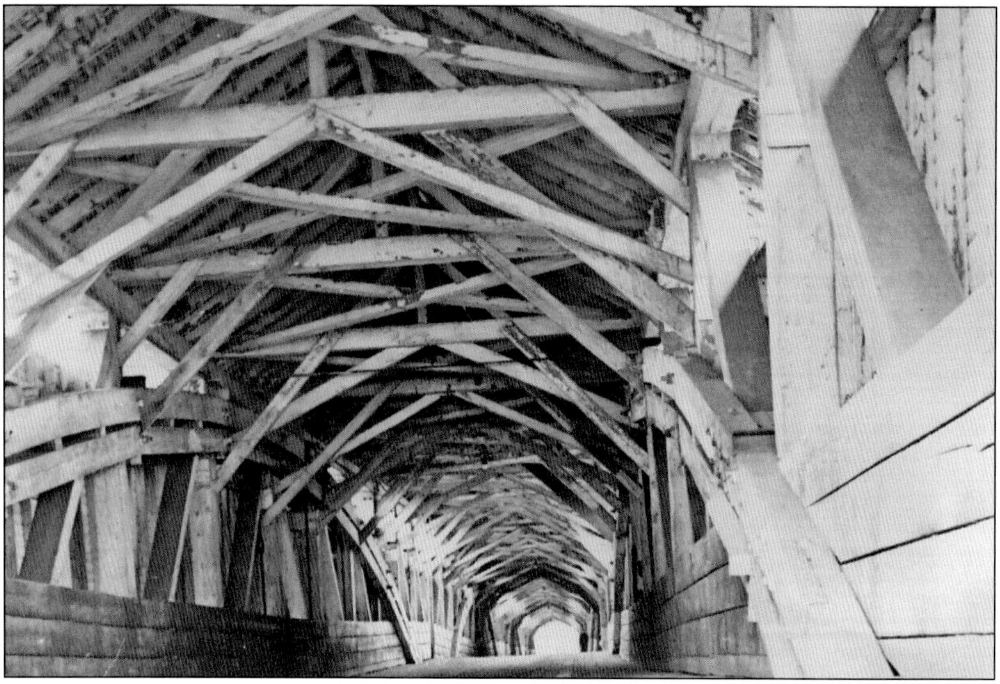

A long financial history preceded the establishment of a covered bridge between Columbia, New Jersey, and Portland, Pennsylvania. The first company was formed in 1817. Another charter was eventually given in 1839, and piers and abutments were built. But not until 1869 did the bridge actually get built. It was a Burr-truss design, at 775 feet long and 18 feet wide. This Portland-Columbia bridge survived the major Pumpkin Flood of 1903 and withstood an actual tornado in 1929. A decision by the bridge's new owners, the Delaware River Joint Commission for the Elimination of Toll Bridges, turned it into a pedestrian bridge (below) in 1953, when a new bridge for automobile traffic was opened just downriver. The last covered bridge over the Delaware River was now literally a footbridge.

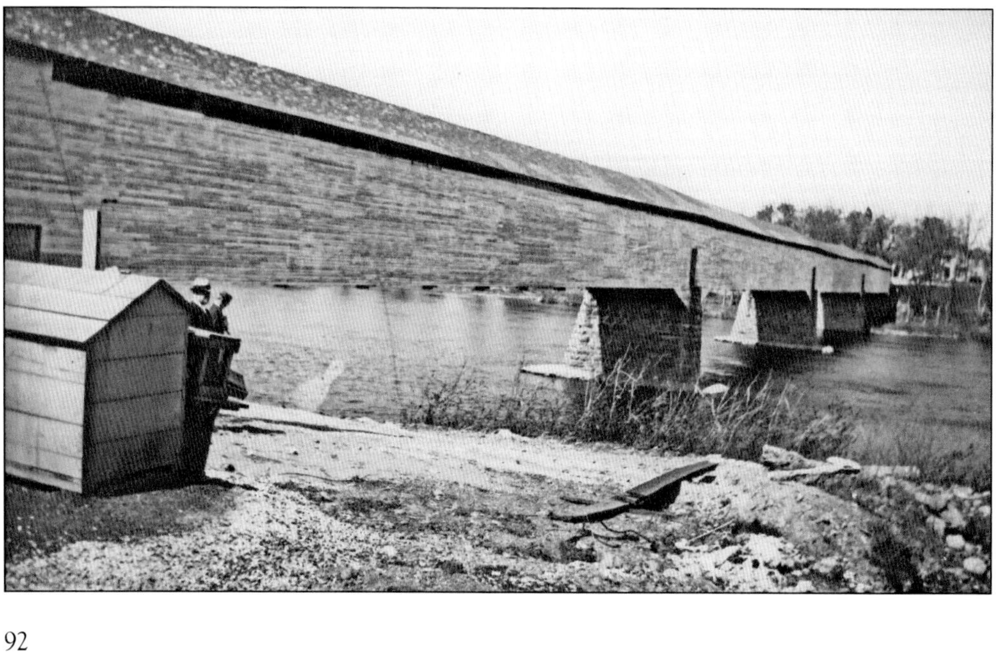

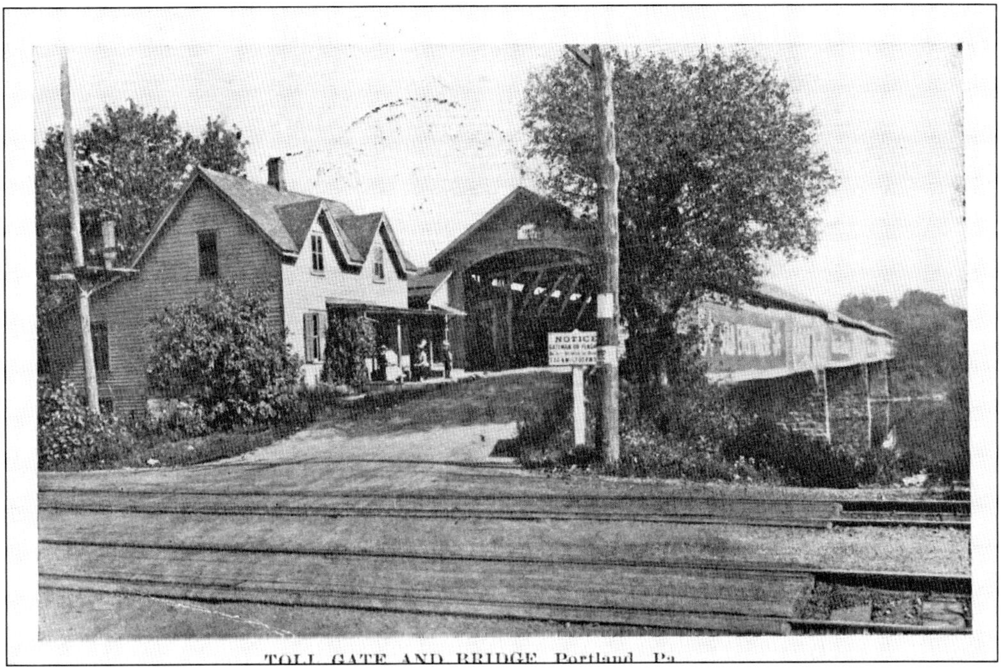

When the first plan to build a bridge over the Delaware River between Columbia, New Jersey, and Portland, Pennsylvania, was presented, a charter was approved in 1817. Not until 1839, however, did actual work begin with the construction of the abutments and piers. The bridge was opened in 1869. Tolls were collected on the Pennsylvania side, and a house for the toll collector (above) was built. The bridge became known as the "Coke Bridge" because of the large Coca-Cola advertisement painted on its side. When the state commission purchased it in 1927, tolls were eliminated, and only a small booth (below) remained where the tollhouse had stood.

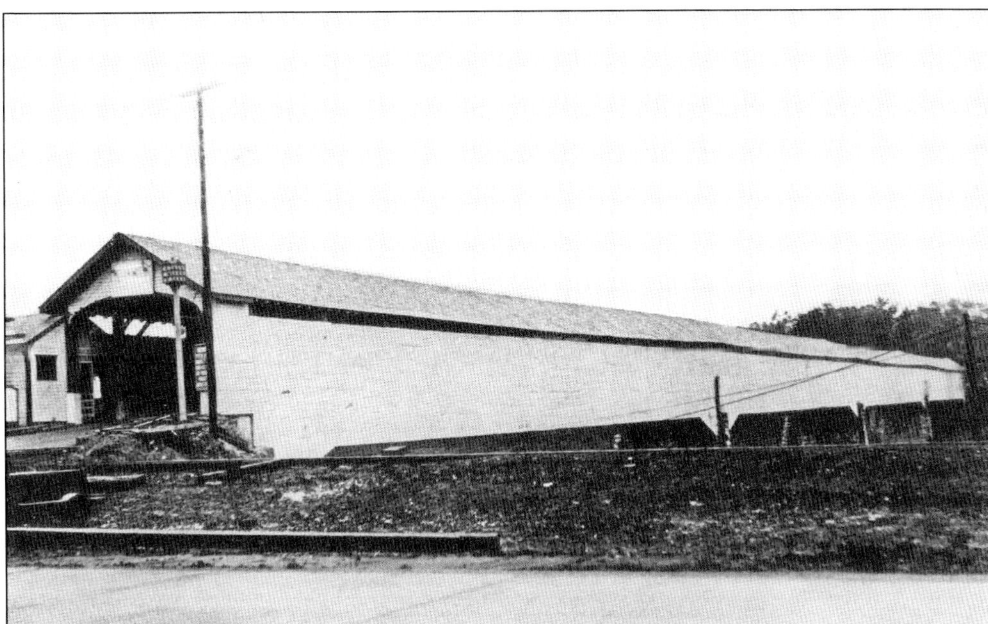

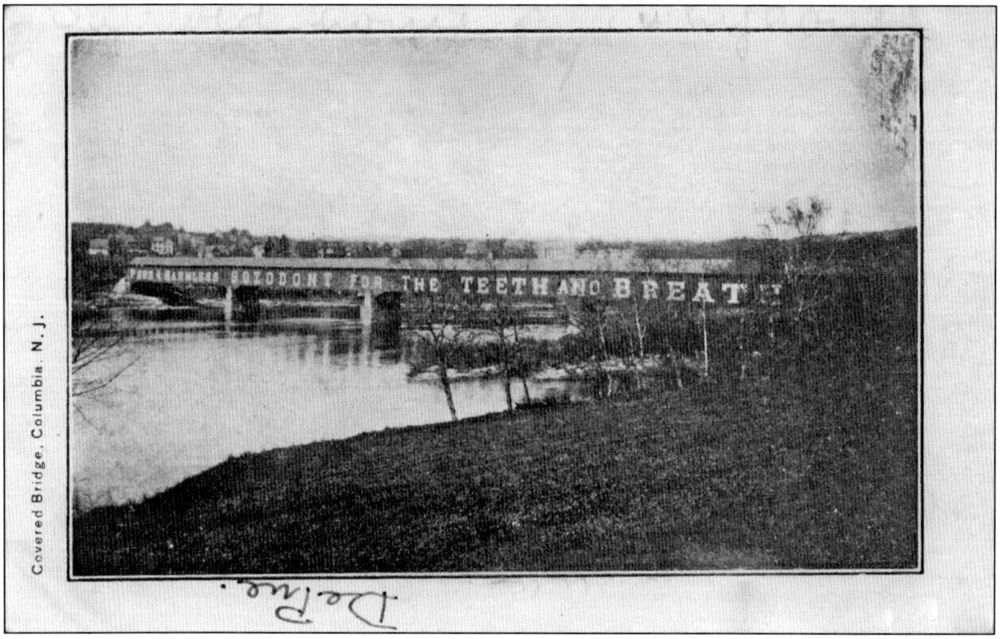

At another time in its history, the covered bridge between Columbia and Portland boasted a very large advertisement stretching across its approximately 775 foot length: "Pure & harmless Sozodont for the teeth and breath." Created in 1859 by druggist Roswell van Buskirk, Sozodont was a popular brand of oral hygiene product from the mid-19th century to the early 20th century.

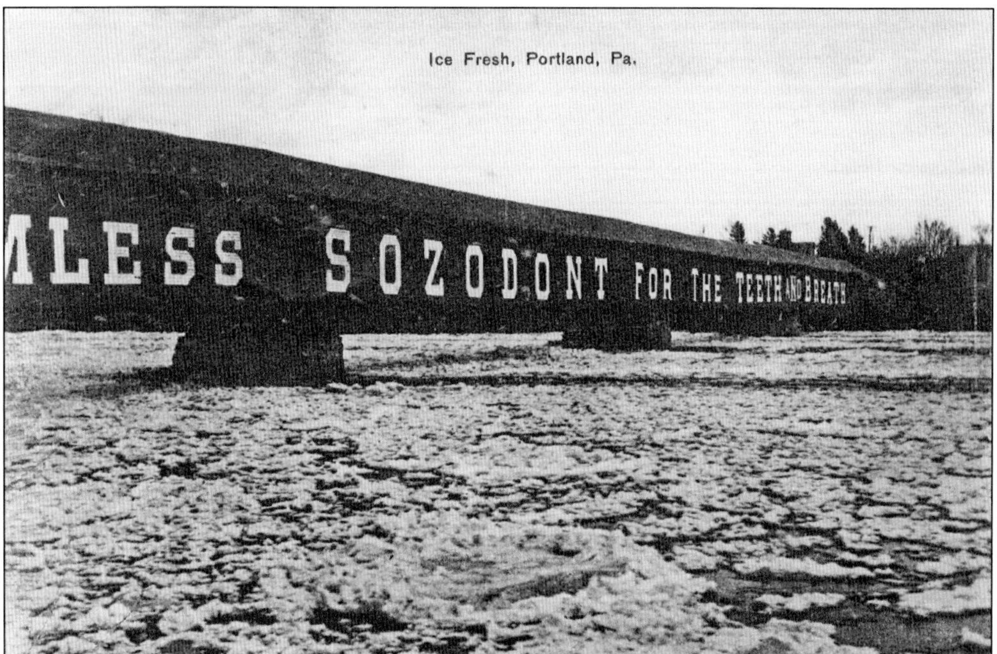

The title of this postcard, "Ice Fresh," is a play on the advertising claim of the product Sozodont, said to "impart a delightfully refreshing taste and feeling to the mouth," with the obvious presence of a river full of ice flowing only a few feet from the bottom of the covered bridge in a spring freshet.

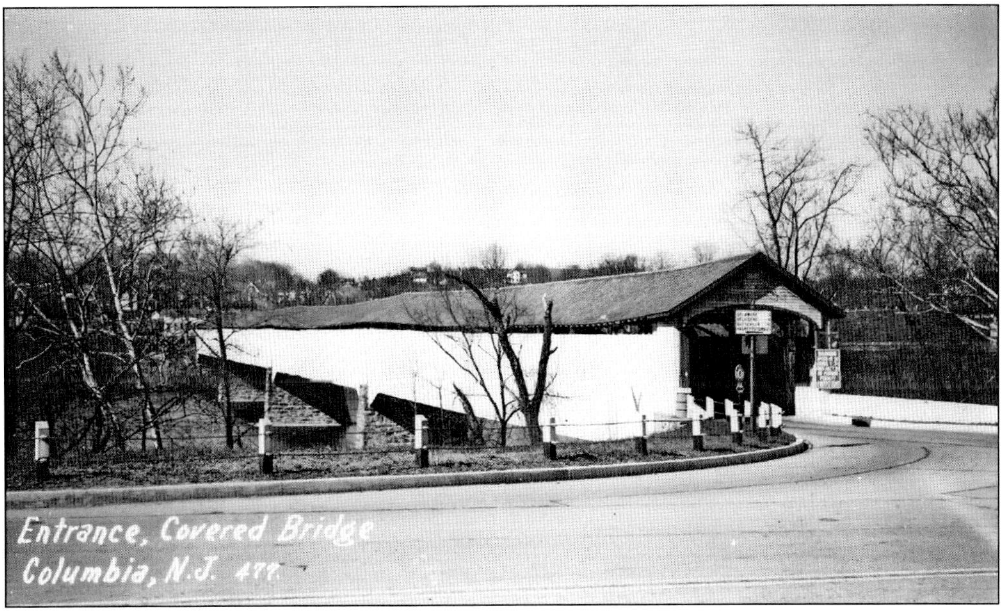

The photograph above shows the New Jersey portal of the Portland-Columbia Covered Bridge in its final time as a vehicular bridge with a 15 mph speed limit before closing to traffic at the end of 1953. The ads were gone, and the bridge wore a crisp coat of white paint. Only pedestrians were allowed to cross. This only lasted a short while. Glen Lewis of njskylands.com recorded an eyewitness who watched the rising, raging floodwaters brought on by Hurricane Diane's torrential rains on August 19, 1955: "The covered bridge could not hold back the river once the flow began to press against the wooden sides." (Above, author's collection; below, courtesy of Todd Clark.)

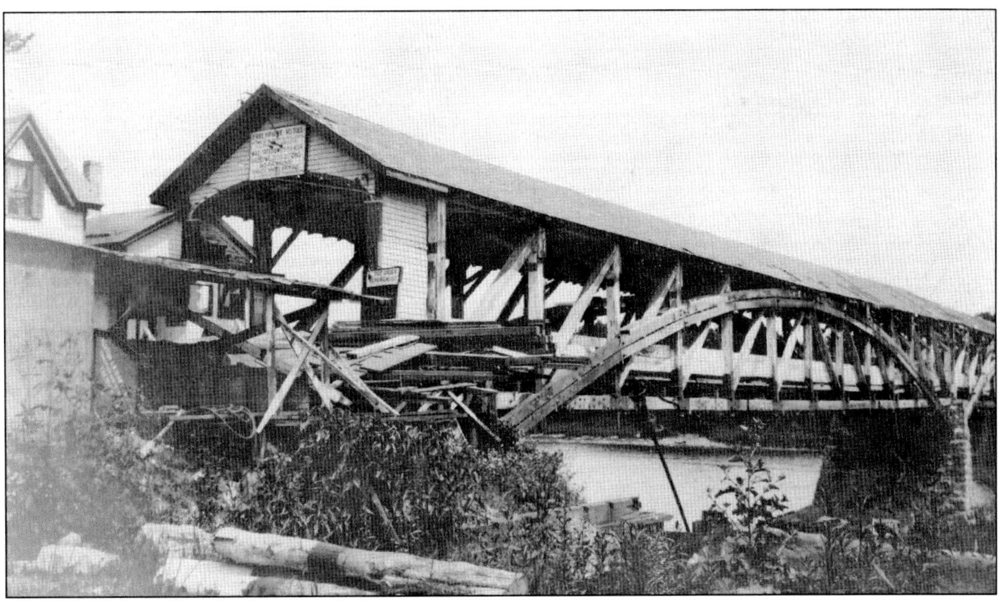

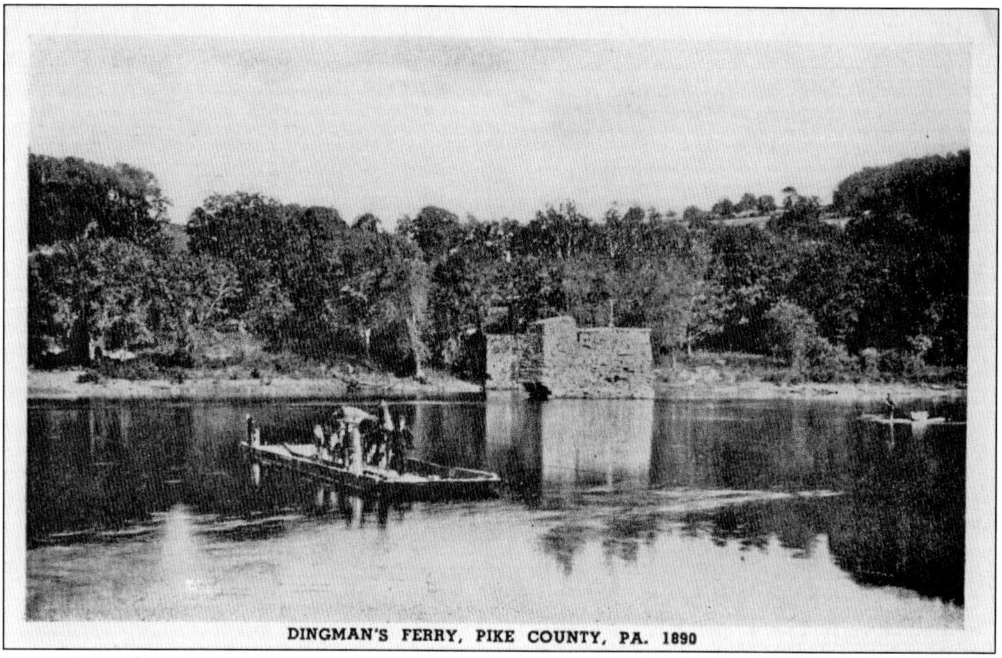

In 1735, Andrew Dingman picked a place on the Delaware River to live, calling it Dingmans Choice. He built a flatboat and operated a ferry. A covered bridge was erected, but high water soon destroyed it. Again, the ferry was called into service. A second covered bridge was blown down in 1865. From 1865 to 1900, Dingman's family operated the ferry.

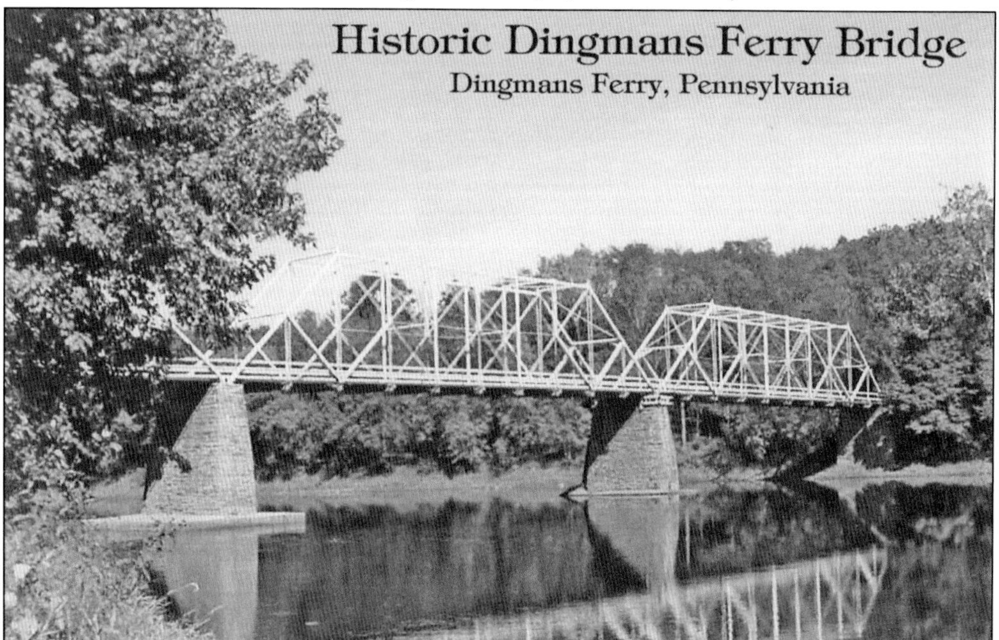

The third and present bridge was erected in 1900. The Dingmans Bridge spans the Delaware River between Dingmans Ferry, Pennsylvania, and Layton, Sussex County. The present company operates under the original charter of 1834. One of the last remaining privately owned bridges in the country, this bridge is a link to more than two centuries of history. (Courtesy of Carol Phillips.)

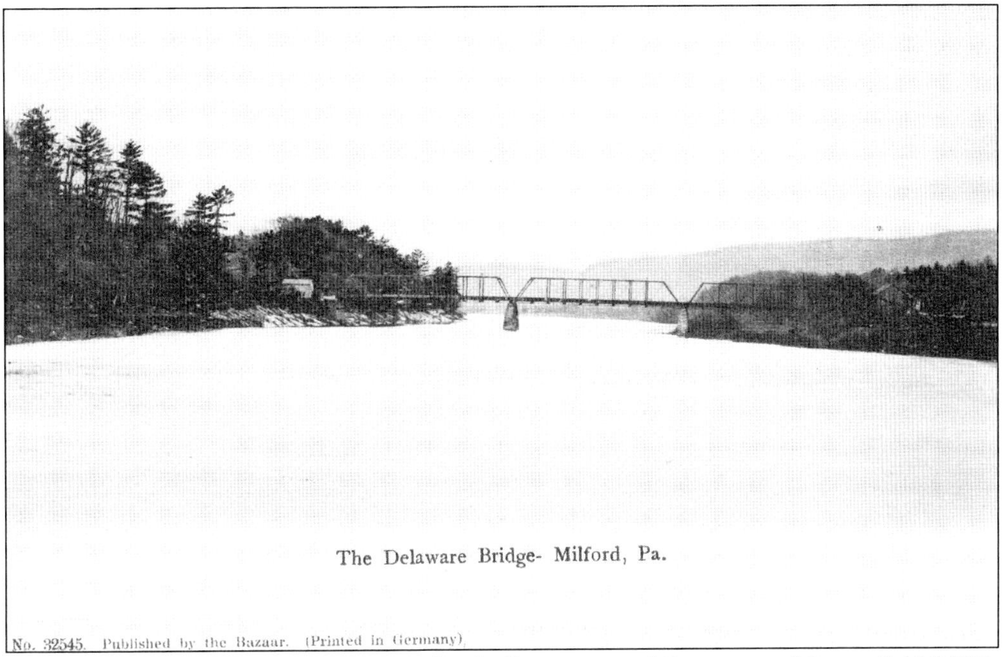

According to the Montague Association for the Restoration of Community History (MARCH), a bridge company was first incorporated in 1804. The bridge, originally a three-span covered bridge of wood construction, 531 feet long, was finally opened for travel in 1836. After a disastrous history, it was again rebuilt of iron in 1889. (Courtesy of John Newcomer.)

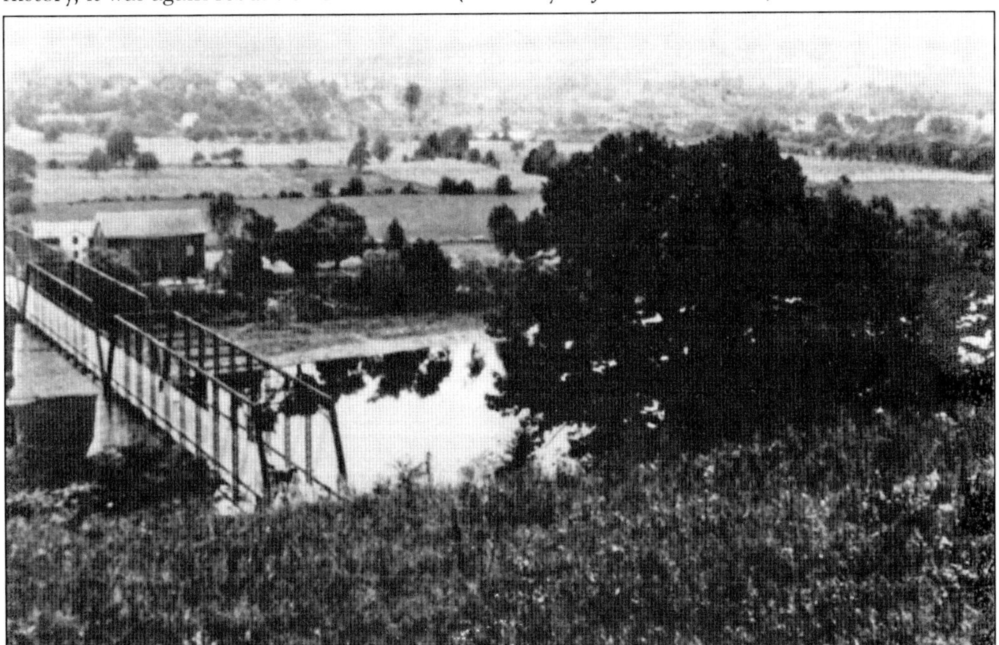

A horse and carriage cross the Delaware–Upper Mount Bethel Bridge at Montague, Sussex County. "Built to take the place of a ferry on the original road from New York to Buffalo, the bridge served its purposes in its day, but now two cars have to pass slowly to avoid side-swiping," as reported in the *Gazette* of Port Jervis, New York, on July 27, 1938. (Courtesy of MARCH.)

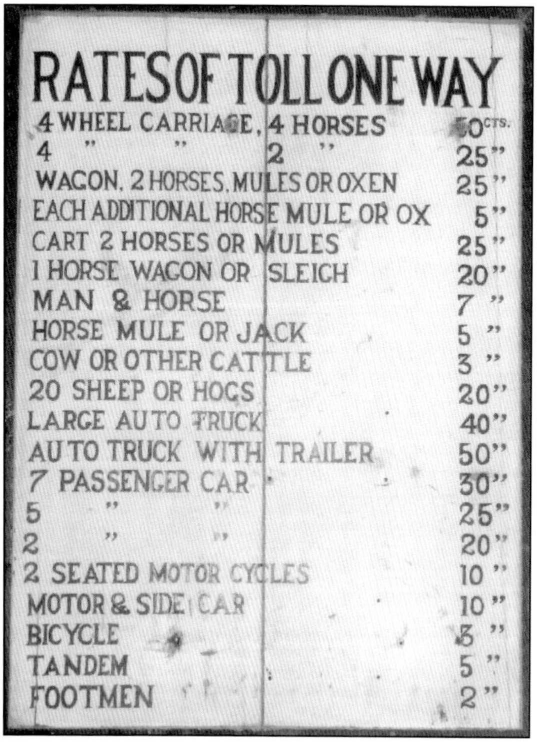

This interesting sign, which was located on the New Jersey side of the Montague Bridge, was typical for all early covered spans crossing the Delaware River. Building a bridge was a business. The company needed to be chartered by both states, and then investors and stockholders were needed. Tolls were the obvious means to pay debt and hopefully provide a profit. (Courtesy of Alicia Batko.)

Men from the Delaware River Joint Bridge Commission were observed while taking a census of vehicular traffic at the Montague Bridge. Many years later, built at a cost of $2 million and overseen by the commission, a modern bridge was opened in early 1954. This bridge survived the disastrous hurricane flood of 1955 and continues in service. (Courtesy of Joseph Donnelly, DRJTBC.)

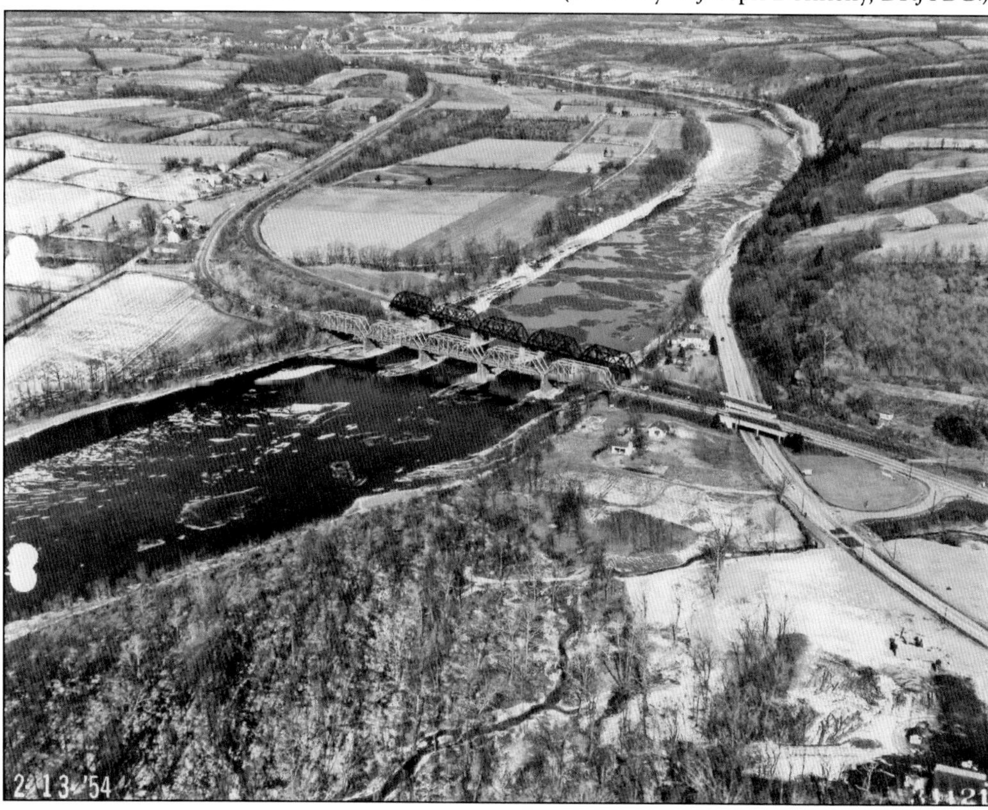

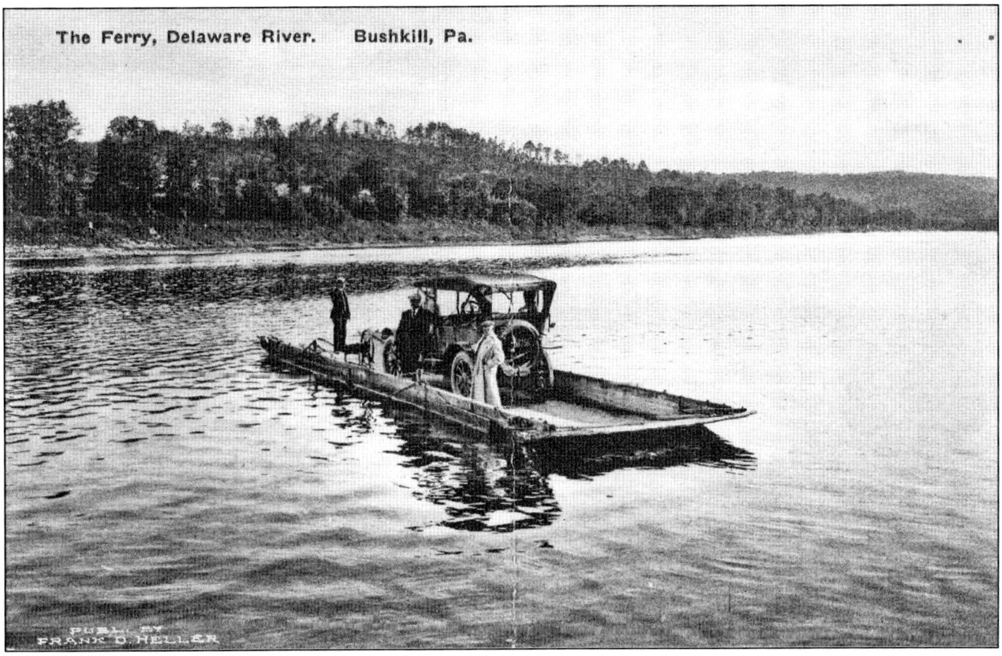

The Walpack Ferry operated between Flatbrookville, New Jersey, and the point of Walpack Bend in Pennsylvania. Between 1744 and 1898, it was also known as Decker's Ferry, Grube's Ferry, and Smith's Ferry. Philip Rosencrans purchased the ferry and, in 1898, moved it upriver to Bushkill, Pennsylvania. As Rosencrans' Ferry, the service lasted until 1945, when a plane accidently cut the cable.

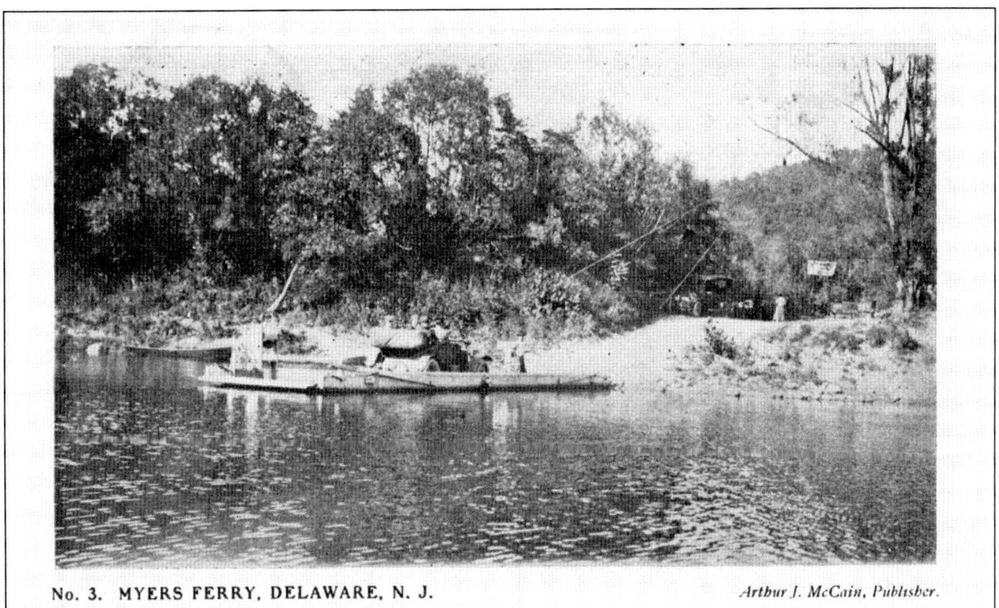

Motor vehicles wait on the riverbank for the Myers Ferry to complete its trip across the Delaware River. This ferry was located at the end of Ferry Lane on the New Jersey side. Begun as Attine's Ferry in 1783, it became Alburtain's Ferry in 1803, Auter's Ferry in 1826, and then Myers' Ferry from 1887 to 1912, but the Myers name continued much longer.

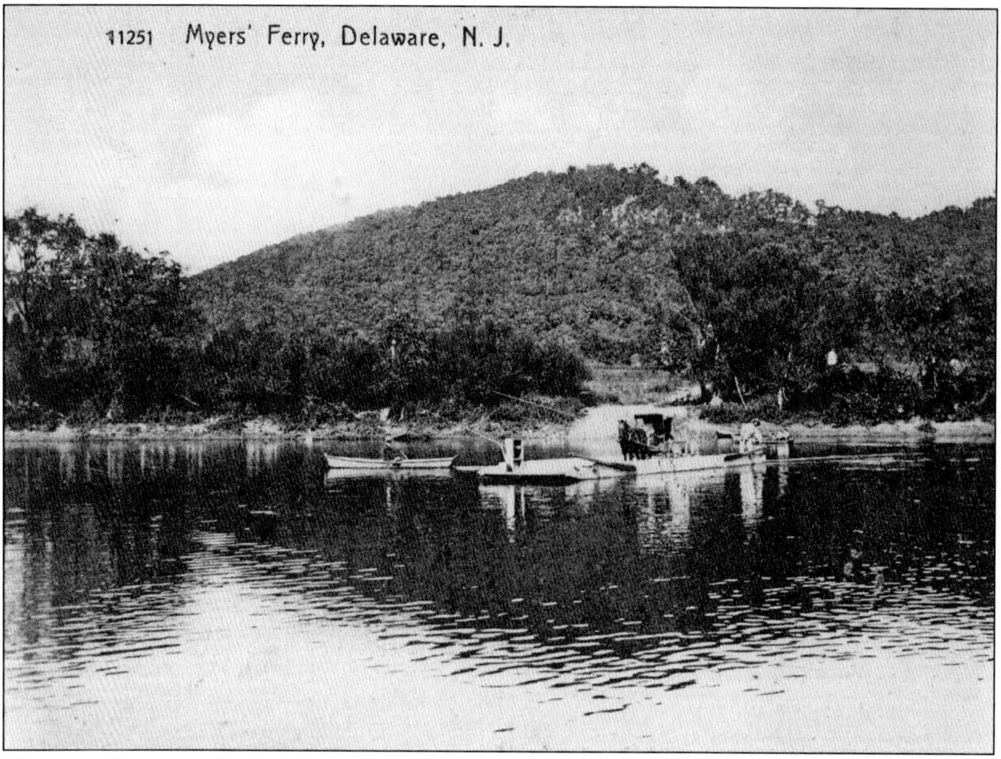

The use of ferries on the Delaware began early in the 1700s. According to Lequear in his *Traditions of Hunterdon*, later ferries were long narrow boats with flat bottoms and vertical sides. The bottom sloped upward at the ends, which had hinged flaps that turned in while crossing and out for the landing. Above, a horse and carriage cross the river. The horse appears calm. Below, three motor vehicles are on one ferry, which was advertised in the *Automobile Bluebook of 1912* as carrying two cars for 25¢ each. Not always a safe venture, a ferry crossing in November 1912 resulted in two automobiles sliding from a ferry when a huge raft of railroad ties struck the boat.

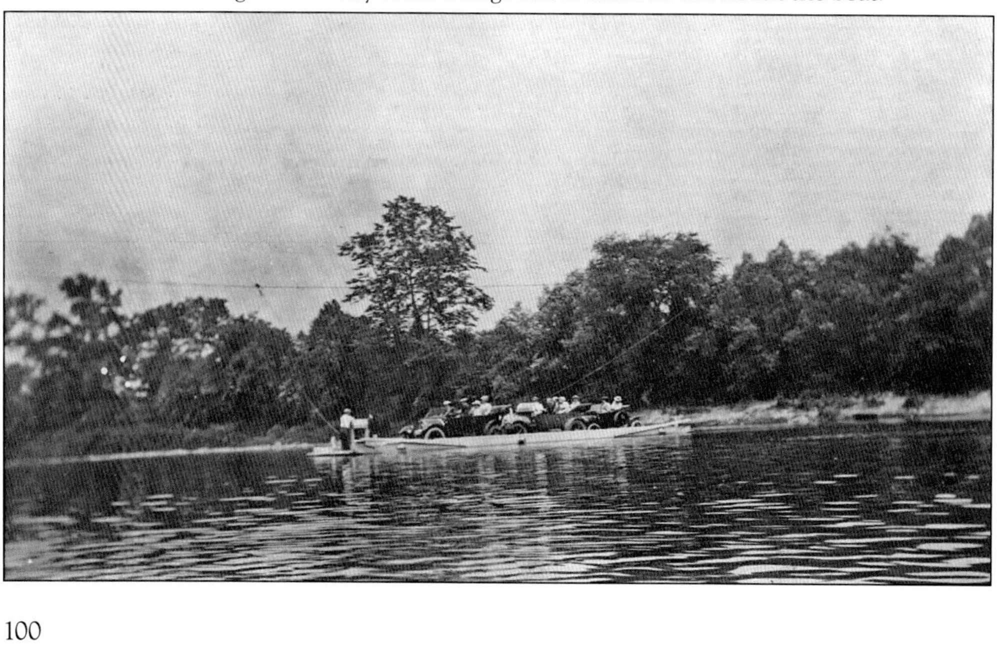

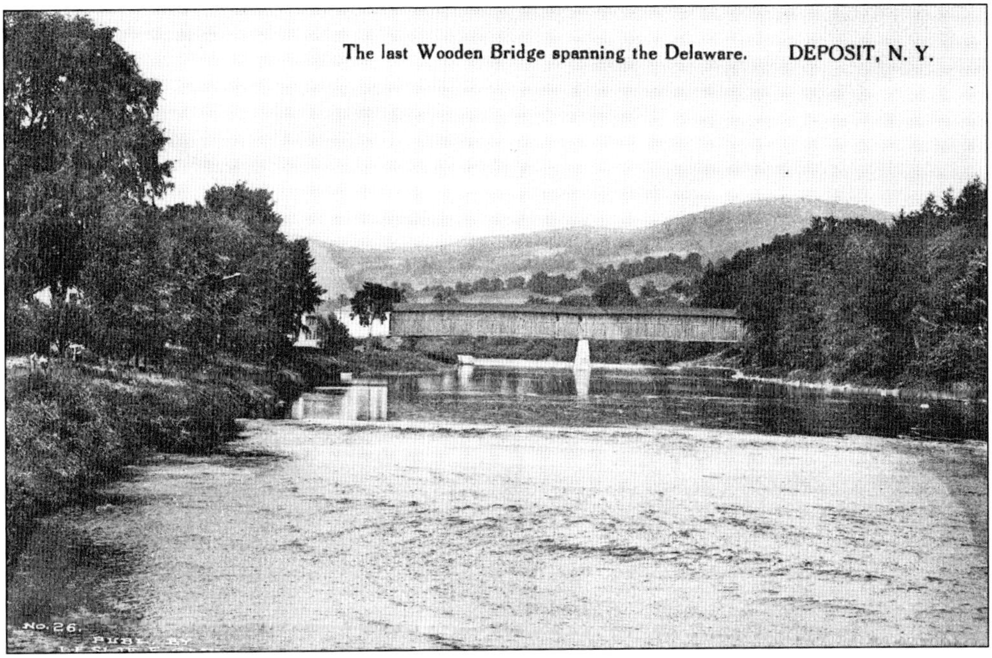

The structures and the histories of the covered bridges crossing the Delaware River between New Jersey and Pennsylvania are venerable, indeed. One final tribute to a wooden covered bridge leads briefly across the New York state line and upriver to a small hamlet called Hale Eddy, where a covered bridge crossed the West Branch of the Delaware River and was called by the name of a larger town five miles north: Deposit. The bridge was built by Robert McMurray and Son in 1865 to replace a wooden bridge destroyed earlier. This covered bridge ably served the area until 1899, when a metal bridge replaced it. The Delaware River again lost another venerable structure.

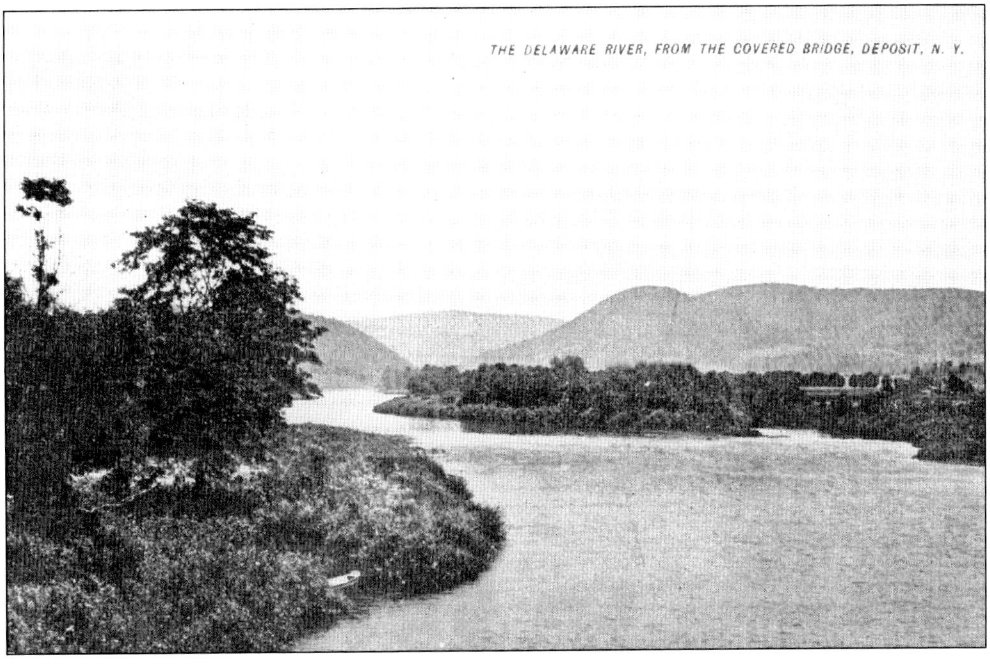

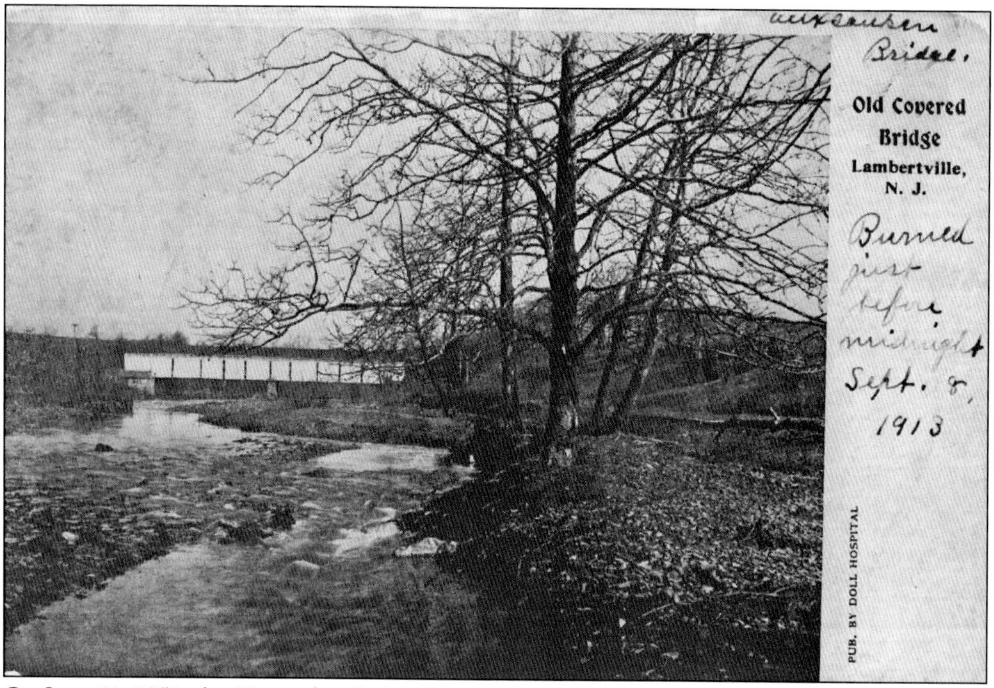

On June 11, 1851, the *Hunterdon County Democrat* reported, "A new bridge had been built over the Aleksaken [*sic*] near Lambertville. It is 170 feet in length and 16 wide in the clear, built upon the lattice plan, resting upon the abutments only, which are 147 feet apart and 3 feet above the August flood [1850] and the greatest rise ever known in the creek." The bridge was again in the news on September 11, 1913, when the *Lambertville Beacon* related, "Fire destroyed the old covered bridge over the Alexanken Creek at a late hour Monday night. When the fire was first discovered by the nearby residents it was seen that both ends were burning. It burned quickly and there was no chance to save it."

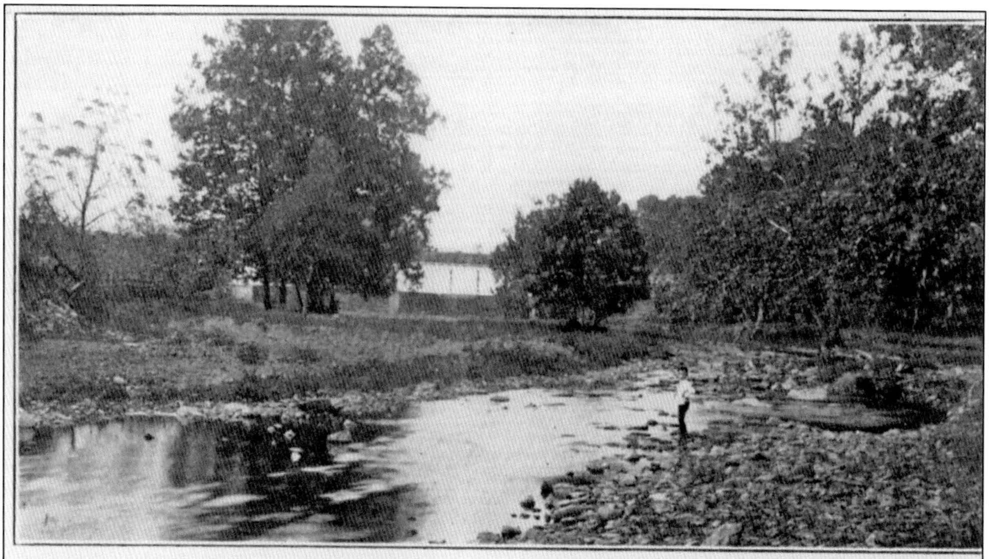

ALEXANKEN CREEK, LAMBERTVILLE, N. J.

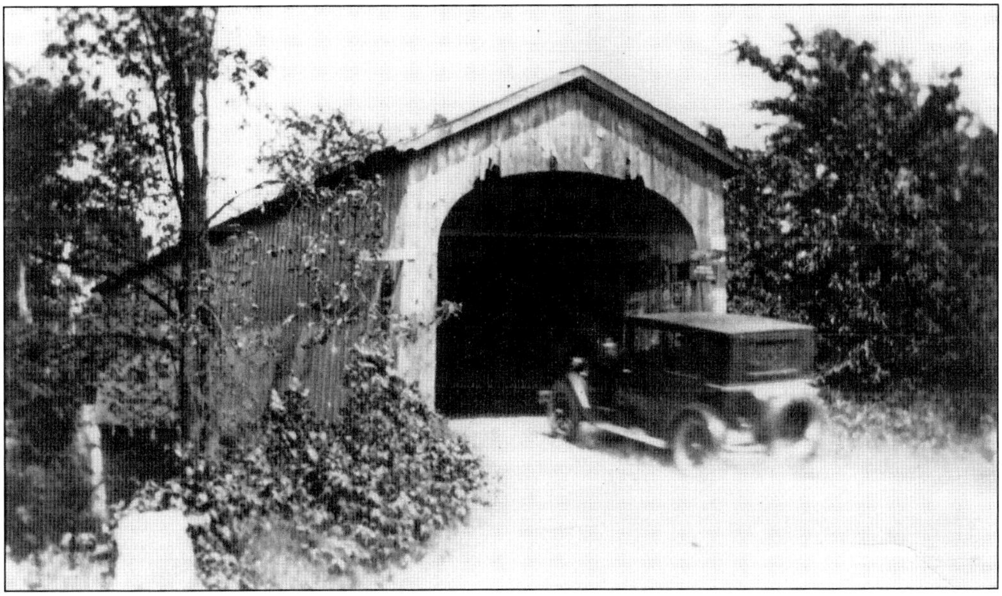

Covered bridges in New Jersey have disappeared due to flood, fire, and progress until only one remains. Although the masonry abutments were built in 1750, the Green Sergeants Covered Bridge was built in 1872. Charles Ogden Holcombe was the designer and chief carpenter using a modified queenpost-truss construction. Of interest in the above photograph is the curved portal framing; pictures including the 1937 Historical American Building Survey report show an angular entry. The bridge is a single span, 85 feet overall, with a 14 foot roadway, vertical planks, an outer width of 18 feet, and a clearance of almost 12 feet. It was rebuilt in 1961 with the addition of steel girders. Painted white, the covered bridge now shares the creek with a stone deck bridge, allowing one-way traffic for both. (Both, courtesy of the New Jersey Historical Society.)

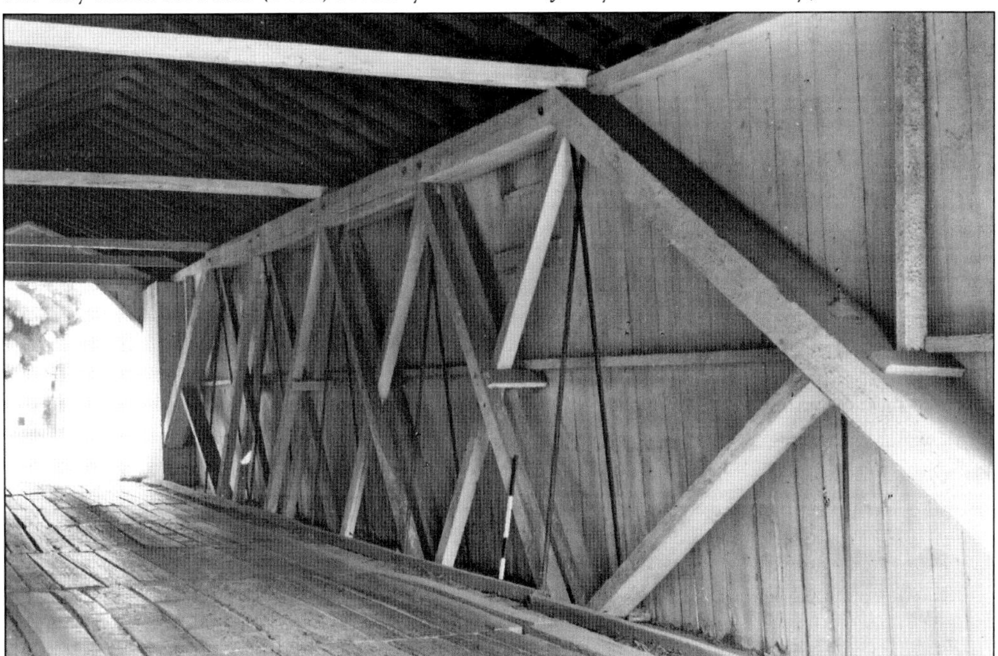

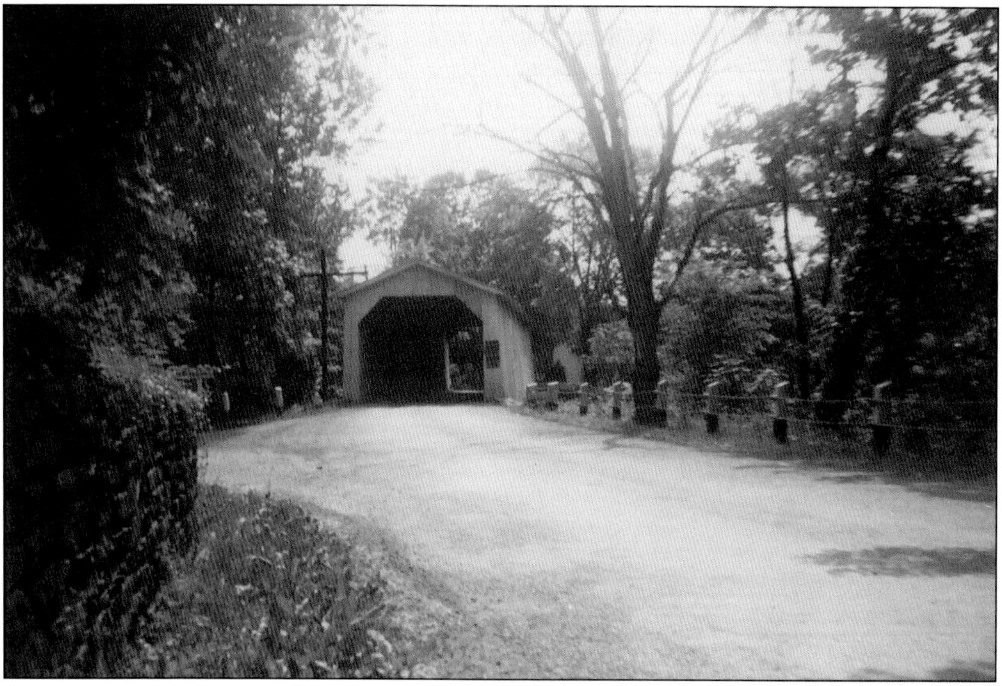

Heading west, one travels a short distance from the town of Sergeantsville in Hunterdon County to cross over the covered bridge. It is said that here, many years before, a tribe of Delaware Indians established a camp after the Iroquois had conquered them. Settled by a Mr. Thatcher in 1700, the village was later named for Charles Sergeant, a Revolutionary soldier.

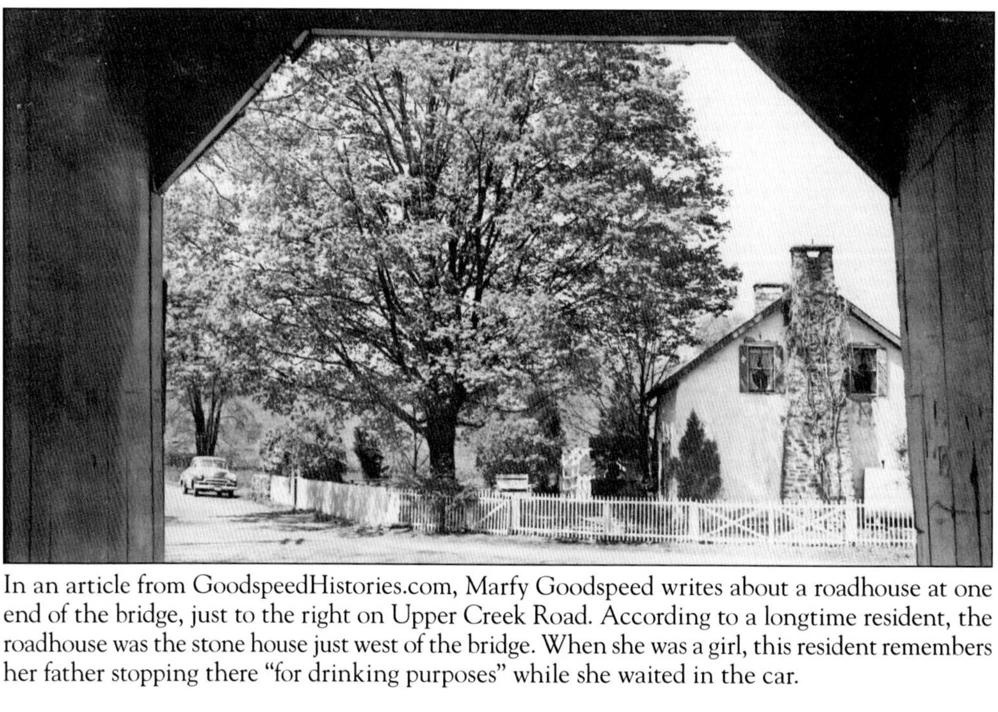

In an article from GoodspeedHistories.com, Marfy Goodspeed writes about a roadhouse at one end of the bridge, just to the right on Upper Creek Road. According to a longtime resident, the roadhouse was the stone house just west of the bridge. When she was a girl, this resident remembers her father stopping there "for drinking purposes" while she waited in the car.

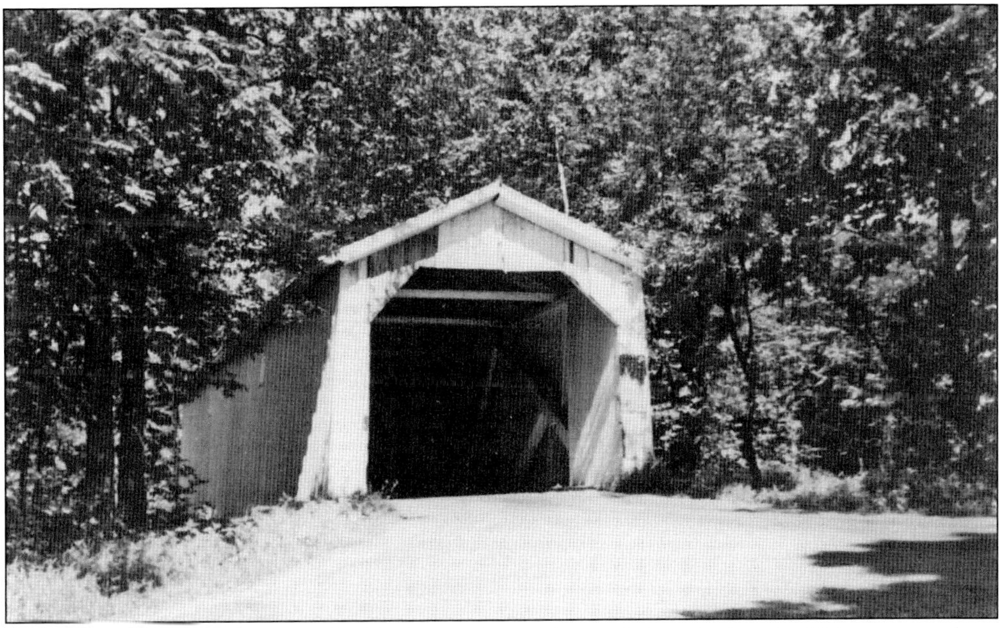

Also on her visit to Sergeantsville, Marfy Goodspeed learned that, during the 19th century, there were heavy wagons carrying stone from the Raven Rock quarries and grain to the mills, but this did not threaten the bridge. In fact, the bridge was such a comfortable, quiet place that weary folk riding their wagons home after an evening imbibing at the Sergeantsville Inn were known to enter the dark bridge late at night and, thinking they had arrived home, stop and fall asleep on the bridge. Sergeantsville was also known as "Skunktown," probably due to a tannery south of town. Skunk pelts were considered attractive and marketable, but the odor near the tannery was extremely pungent.

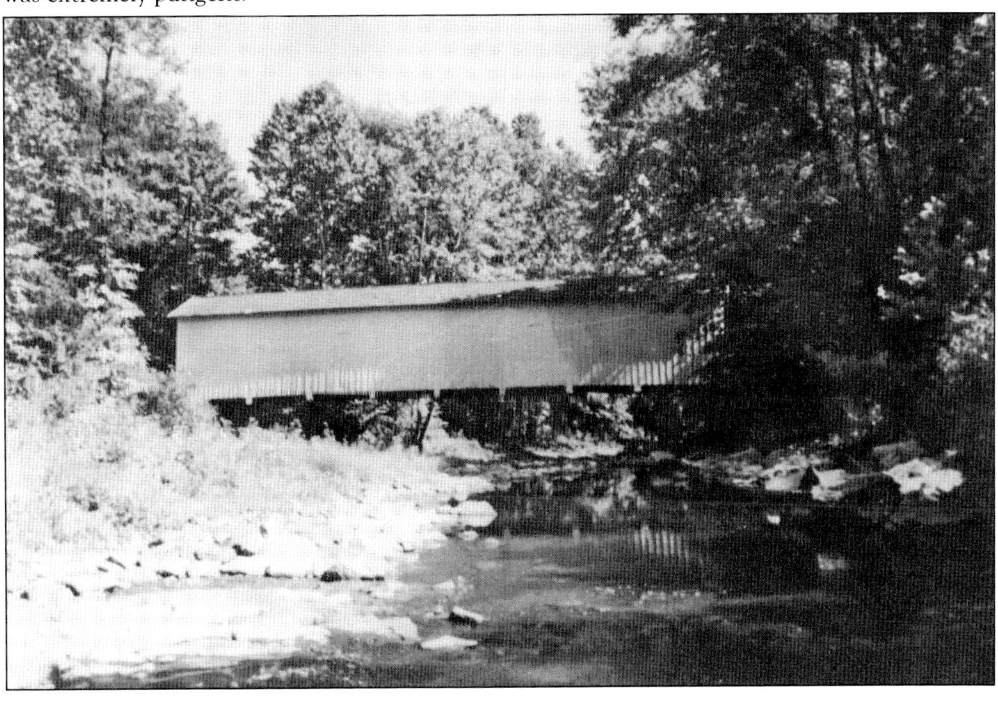

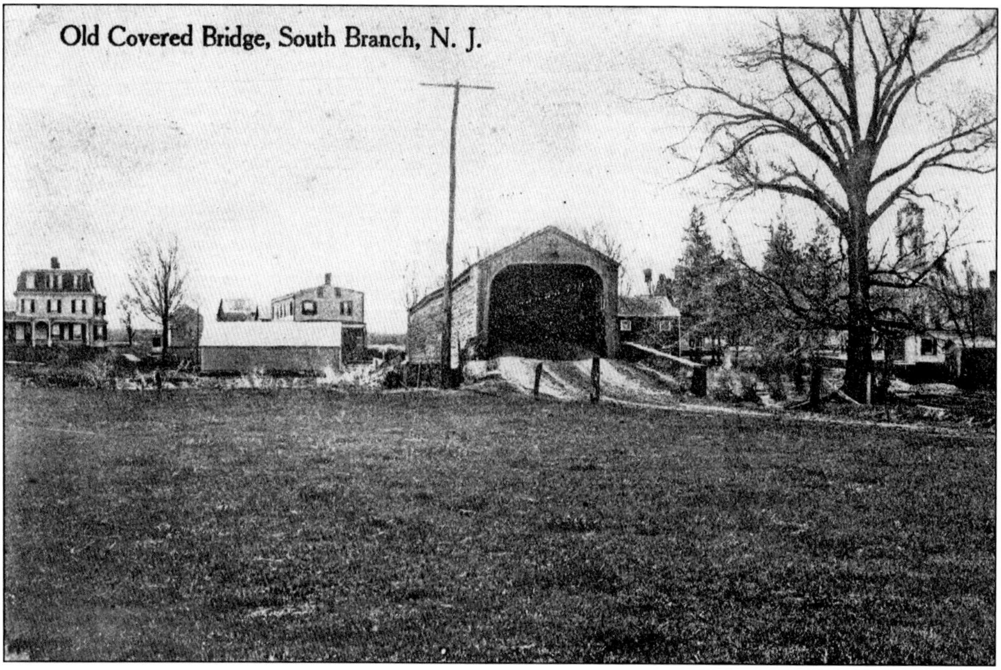

According to *A Brief History of Hillsborough*, Joachim G. Quick built the South Branch Covered Bridge, Somerset County, in 1830 with wood salvaged from the Millstone Reformed Church, which was torn down in 1828. The bridge was built on the Town lattice plan, with trusses as long as 30 feet crafted from half a dozen mortised slabs. Dowels were used to hold the timbers together. It was two spans over the South Branch Raritan River. The first crossing at this site dates back to Revolutionary times. Old residents remember that the children would climb under the bridge in search of bass and would return muddy but joyous at capturing the "big one."

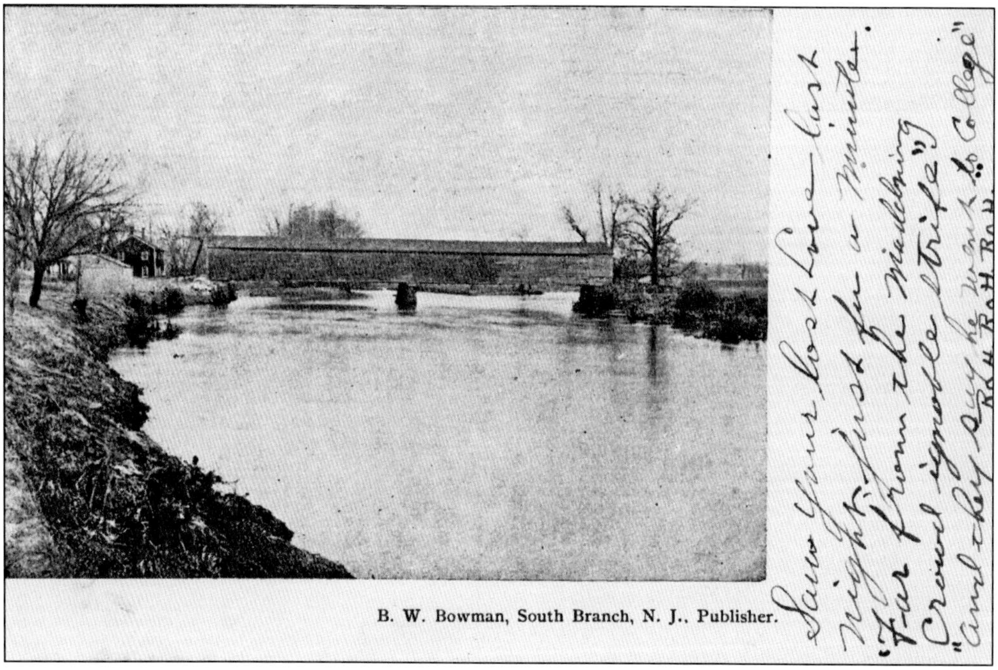

Typically, one car waits as another completes its passage through the Montgomery Covered Bridge, which was located in Hillsborough, Somerset County. The bridge crossed a branch of the Neshanic River, a tributary of the South Branch Raritan River. A single span, the bridge's truss type is uncertain. Built in the 1850s, the bridge was replaced in 1928. (Courtesy of the New Jersey Historical Society.)

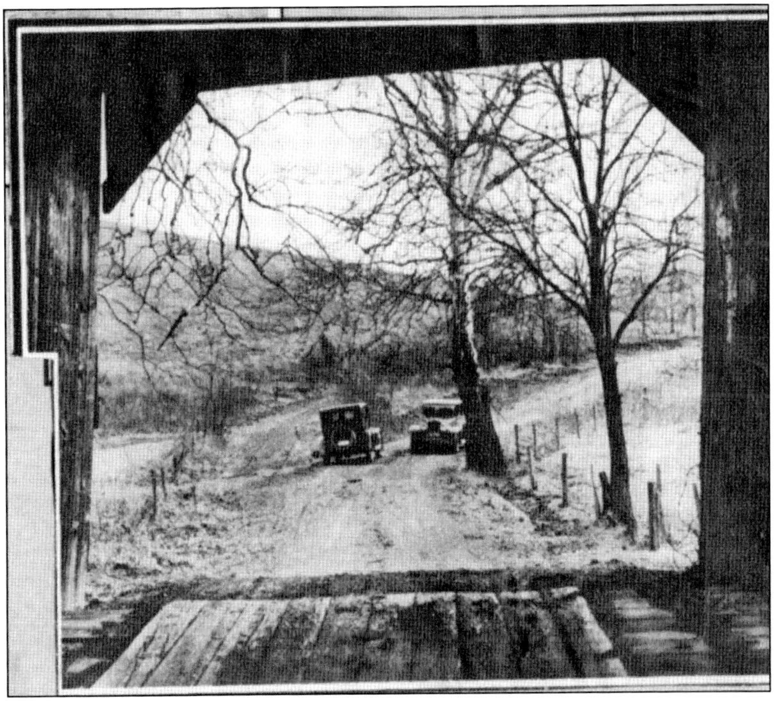

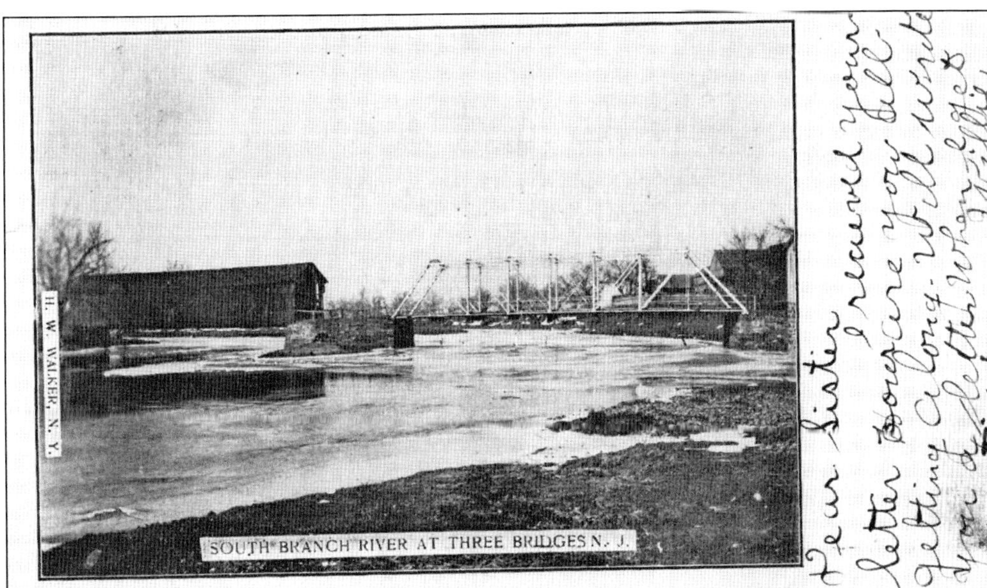

Three Bridges is located on the South Branch Raritan River, Hunterdon County. According to Richard Sanders Allen, two of the three bridges were once covered. When the village applied for a post office, a judge ruled that a name change was out of order and that Three Bridges was to be the name. In 1881, all three bridges were damaged by a severe storm.

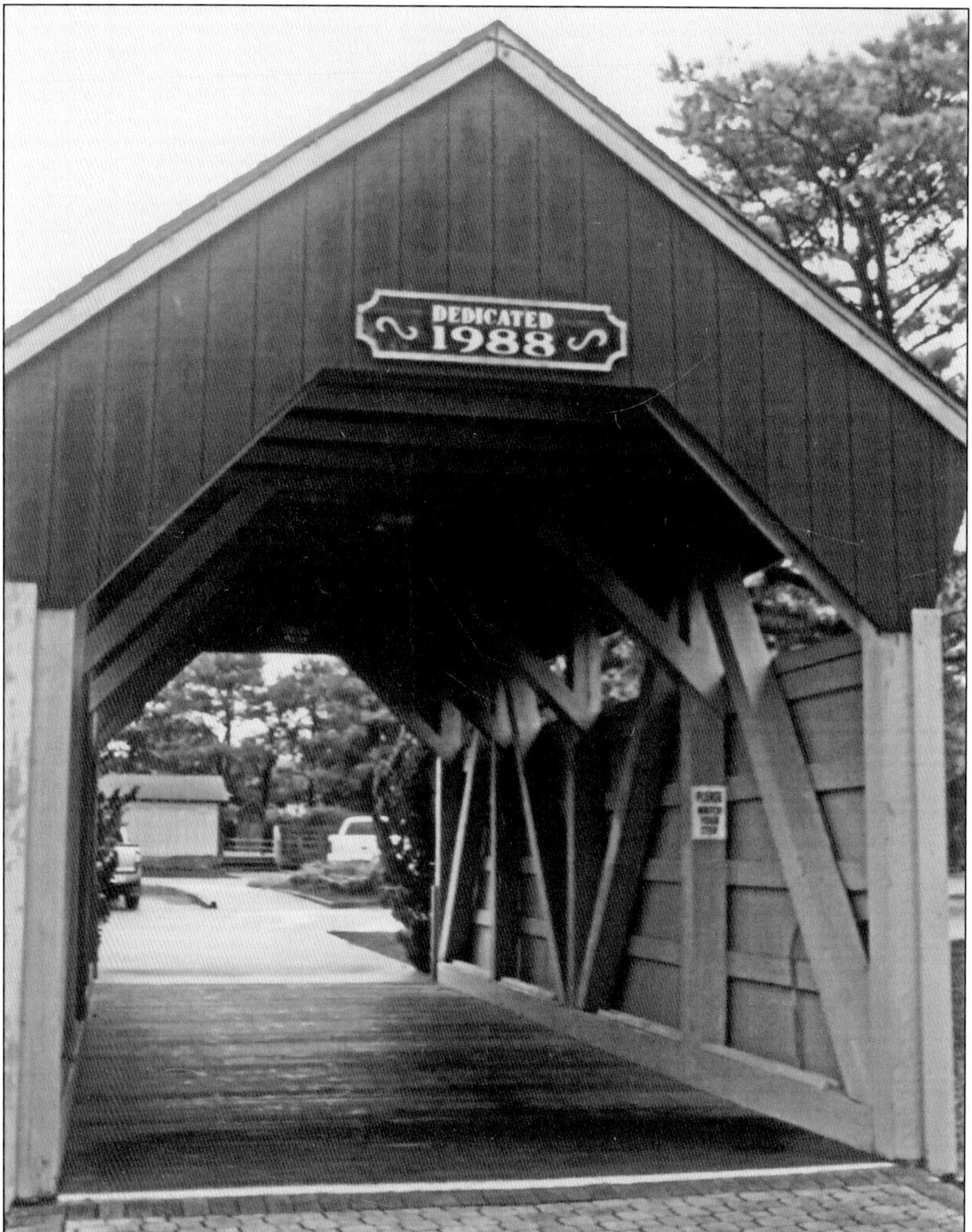

The Smithville Covered Bridge has been in place for 25 years. It was rebuilt in an old historic area. An Englishman, Daniel Leeds, mapped out the land known as Egg Harbor, named for the variety and quantity of bird eggs found in this part of South Jersey. The tidal salt meadow and marsh, interspersed with shallow coves and bays, provided a resting and feeding habitat for water birds on their migration to and from their winter and summer habitats. Long before the Revolution, seamen also found the Little Egg Harbor River (now known as the Mullica River, just a few miles north of the Towne of Historic Smithville) was a perfect place to smuggle goods to the New World. The Little Egg Harbor Inlet and River, which contained a number of sandbars and inland waterways, was a perfect place to auction, barter, and sell their illegal bounty in Atlantic County.

Six

Other Bridges and Their Stories

New Jersey's covered bridges have proven to be both an important factor in the development of the state and an interesting element of its history. In more recent years, some "romantic shelters" have been built with stories to tell. A few other unusual bridges also deserve mention as well. On occasion, there is an example of history repeating itself, as the following narrative by Chris Darlington relates, concerning the Pedricktown Covered Bridge, courtesy of the Salem County Historical Society:

> My paternal grandfather, William Barlow Darlington, was one of the prosperous farmers of southern New Jersey. He established himself in 1905 on a 133-acre farm located on Beaver Creek, a small tributary of Oldmans Creek. My grandfather was prosperous because of his loving care of the soil, the crops he chose to grow and market, and his innate frugality.
>
> In 1915, Grandfather purchased one of the first automobiles in the township. Being well aware of the increasing need for improved roads, the county rebuilt the main road, and a two-lane concrete causeway replaced the old covered bridge. Grandfather acquired the bridge and used the timbers to build a wagon shed. On nearly every visit to the farm, I would head for that shed to admire the stout and powerful timbers which supported its walls and roof. In my imagination, the bridge timbers represented a set of Lincoln Logs held together by large pegs.
>
> My grandfather died in 1941, and years later, my uncle offered the covered bridge timbers to the developer of Smithville Village, a commercial "village" north of Atlantic City. I visited Smithville Village several times, came upon a pile of stout timbers, but gathered from the look of things the bridge would never be built. But I was wrong.
>
> Four or five years ago, I visited Smithville again. There in the midst of this village stood the old covered bridge. It was a real pleasure for me to walk over the old bridge, to feel the stout timbers again, and to appreciate the power of the structure. It is very much a piece of my own history.

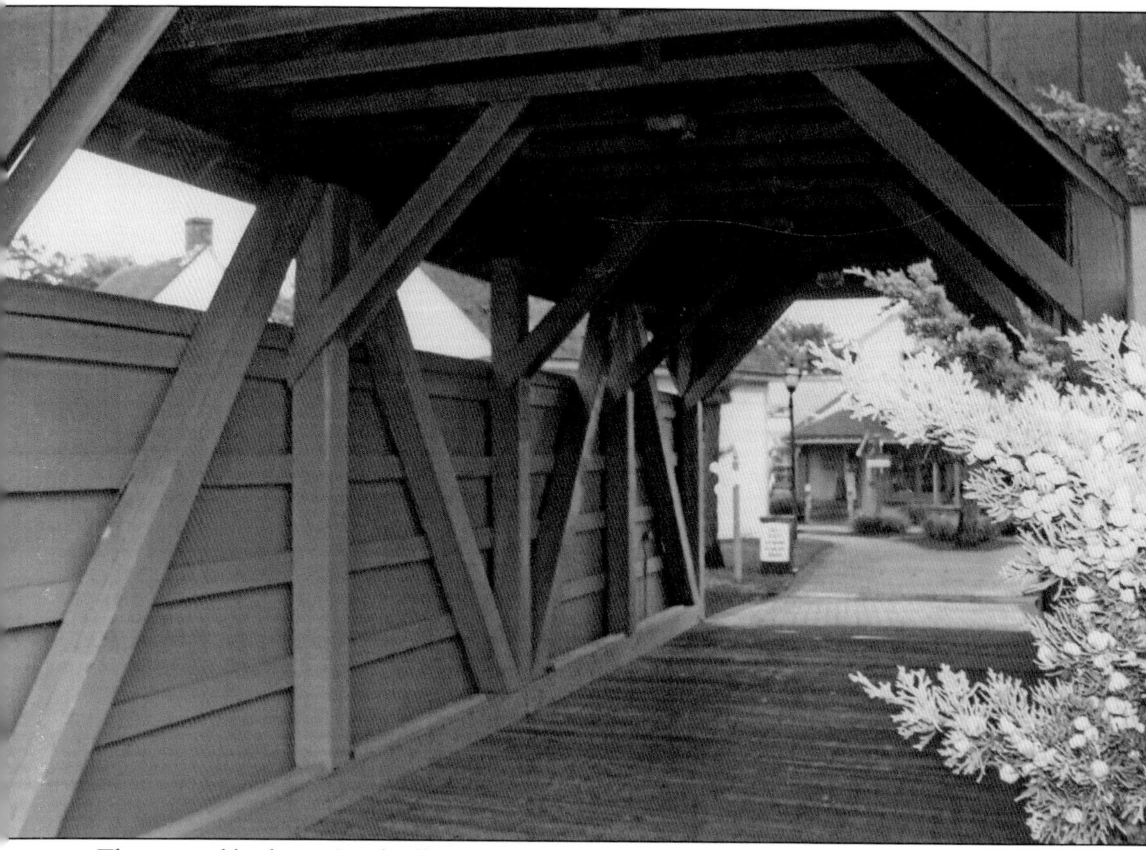

The covered bridge at Smithville, a romantic shelter, was reconstructed and dedicated in 1988. As stated in the chapter introduction, the heavy timbers were taken from the covered bridge at Pedricktown, where their second use became a wagon-shed structure. Now appearing as a double kingpost, the trusses no longer support the bridge. A concrete deck beneath the bridge and lights have been installed, and only foot traffic is permitted. The Smithville Covered Bridge is painted a vibrant barn red. The width of the bridge is approximately 15 feet. The vertical floor planking is new, and the roof rafters are made of new lumber. The original timbers from Pedricktown are painted light beige. Despite being painted, which has seemed to protect these aging timbers, the old truss work lends a positive note to the nostalgia of walking through this bridge.

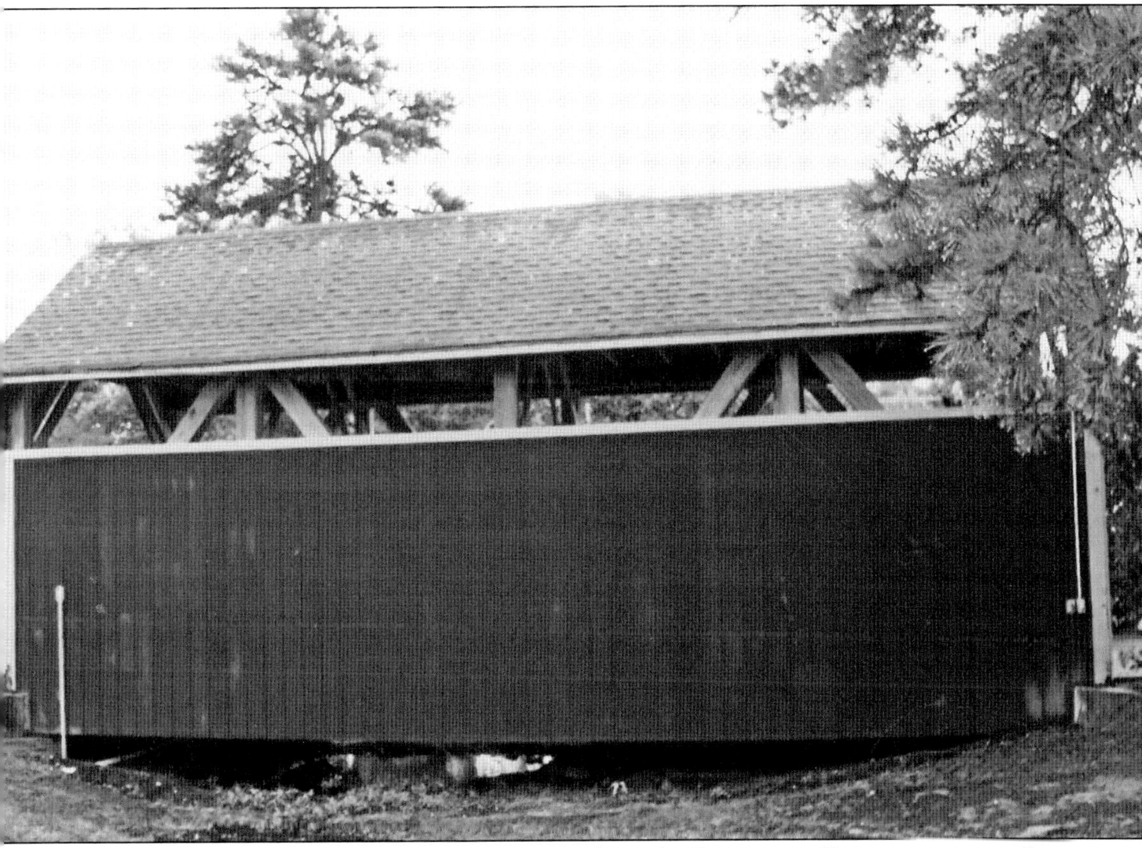

The Smithville Covered Bridge serves as passage from the Barn, a small country inn, to the Village Greene of the Towne of Historic Smithville. In colonial times, the name for this area was Galloway Crossroads. The Village Greene is a grouping of shops and eateries, most located in historic buildings moved to the site. A gristmill is located on the lake. In its previous location, the local farmers would bring to the mill the daily quota of wheat, buckwheat, rye, and oats to be ground. Also of interest, the Jersey Devil, sometimes referred to as the Leeds Devil, is said to have roamed about in the swamps a few miles from Smithville in the direction of the bay at Leeds Point. According to William McMahon, the appearance of the Jersey Devil has been reported from time to time through the years, yet he never seems to age.

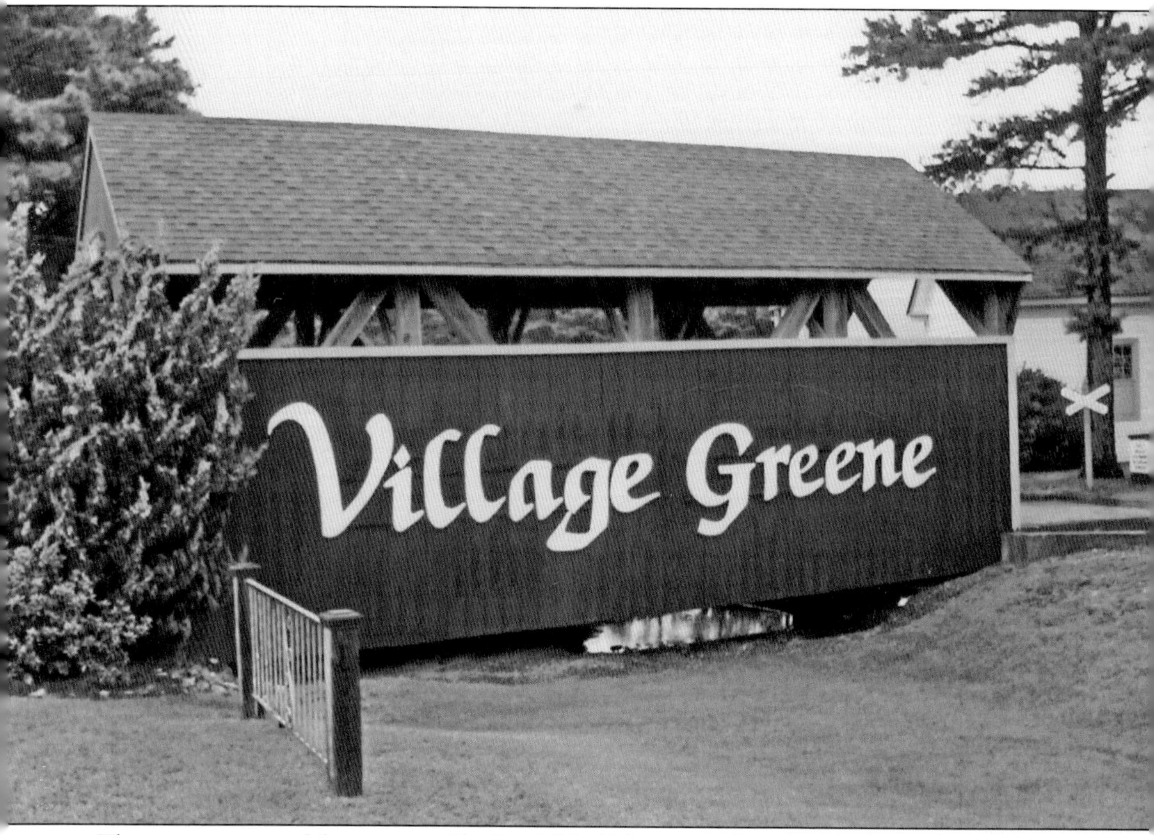

Then new owners of the property, Fred and Ethel Noyes, purchased and relocated many historic buildings during the 1950s and 1960s. They searched throughout South Jersey to find and obtain these buildings and bring them to Smithville. In 1964, the Towne of Historic Smithville was named a National Historic Site. In 1974, the Noyes family sold the town, and the property changed hands several times, actually ending up at a sheriff's auction. Now, a partnership owns the east side of Lake Meone, Historic Smithville, and Ed and Wendie Fitzgerald purchased the west side of the lake, the Village Greene, which includes the covered bridge, continuing this rich American tradition. Smithville provides a quiet and relaxing atmosphere away from the hustle and bustle.

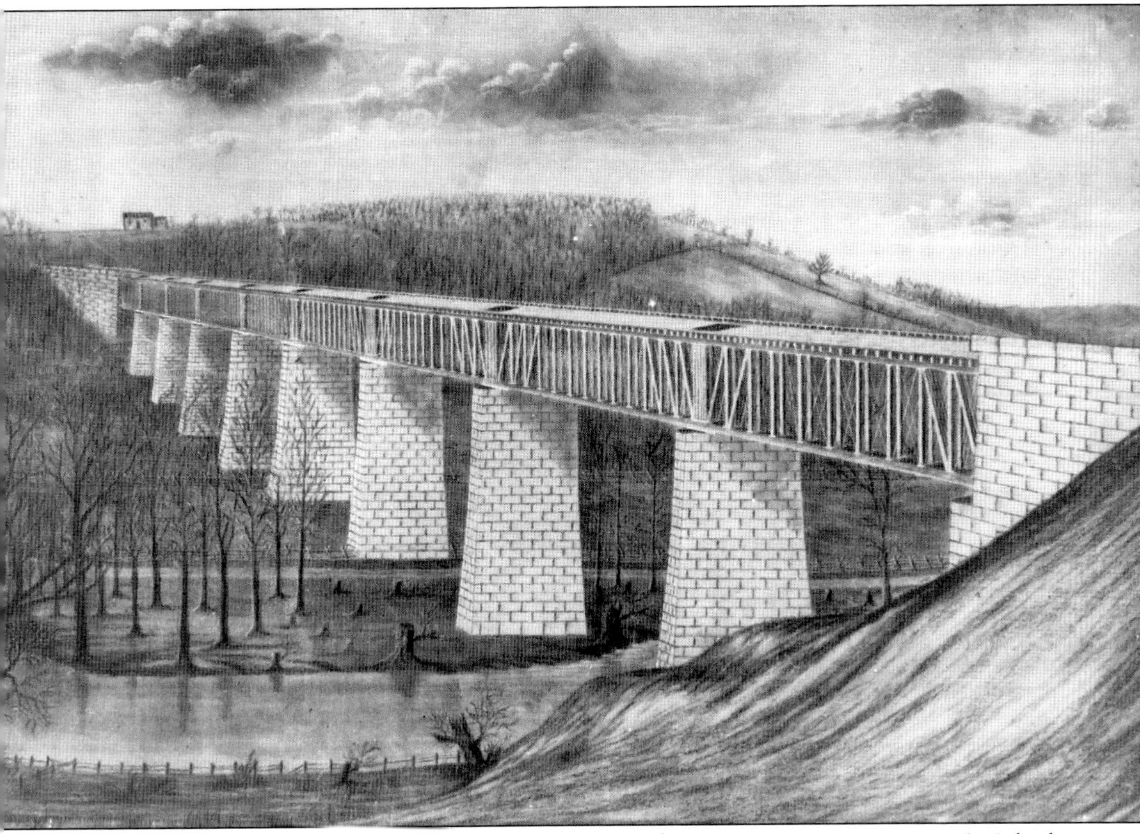

The Central Railroad built this railroad bridge at High Bridge, Hunterdon County, in 1852. It had a deck-type truss of uncertain type. William Umek found in the *Guidebook of the Central Railroad of New Jersey* that the bridge had nine spans on eight tall piers and was 105 feet high and 1,300 feet long. The *Guidebook* continues, "But the glory of this bridge has now departed; it has lost its old charm of picturesqueness. The railroad company has for three or four years been transforming it into a lofty embankment with a double arch culvert." The High Bridge Branch connected to the Delaware, Lackawanna & Western Railroad, the Wharton & Northern Railroad, and the Mount Hope Mineral Railroad. Most of the line was abandoned soon after Conrail assumed operations in 1976. On November 1, 1906, a Flemington newspaper published this picture of the old railroad bridge. (Courtesy of the Hunterdon County Historical Society.)

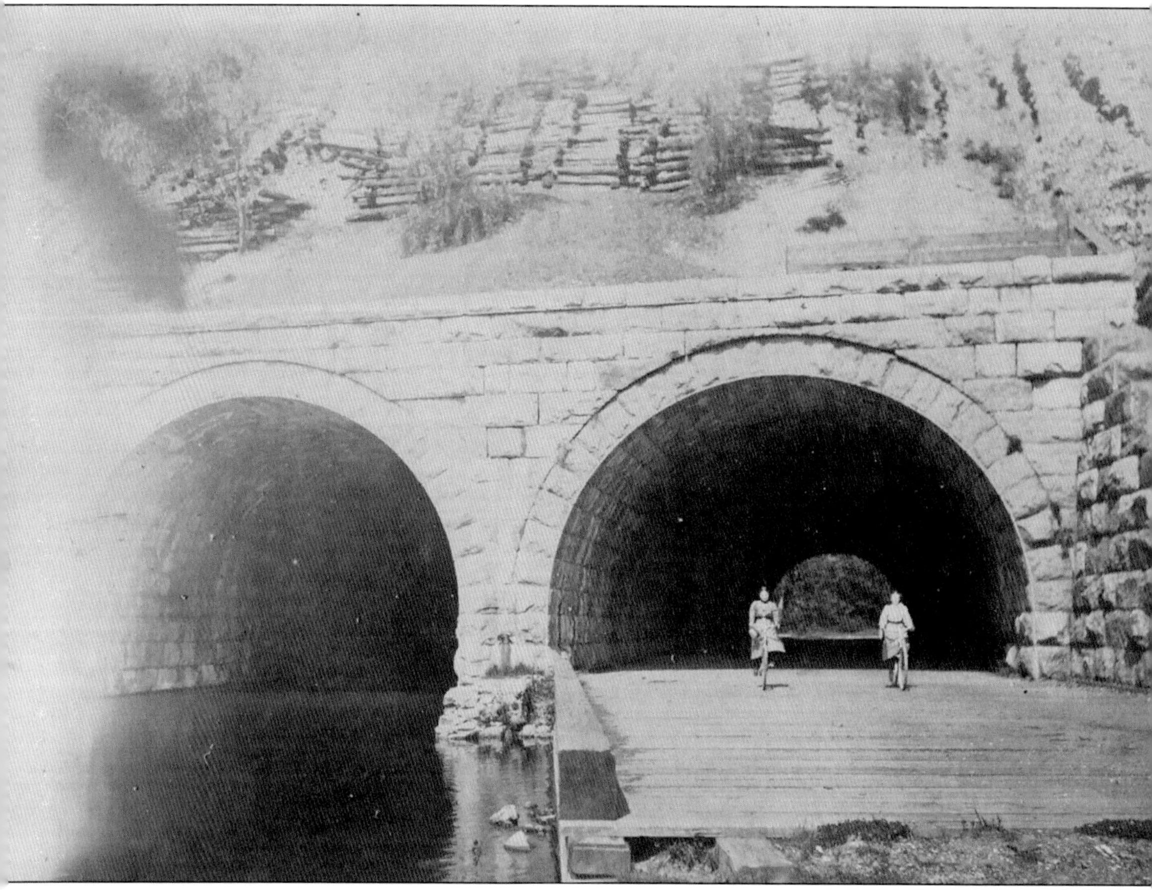

The Arches are a result of the weakened and dangerous conditions of the High Bridge originally built over the South Branch Raritan River at this location. The town had already become known as High Bridge because of this structure built in 1852. Researched by Douglas Kiovsky of the Hunterdon County Parks Department, he wrote, "After almost a decade in use, in 1859, it was determined that the railroad's high bridge had to be replaced. The weight of the trains between the stone piers of the bridge caused it to sag creating a dangerous roller coaster effect, but in reality burning embers from the coal engines would land on the wooden trestle potentially causing a fire swept by the winds of the valley floor. As a precaution the bridge site and the stone piers were buried with fill by 1865. Two arching stone culverts were built at the base of the bridge, one for the river and one for safe passage for horses and carriages." This arch is called Arch Street. (Courtesy of the Hunterdon County Historical Society.)

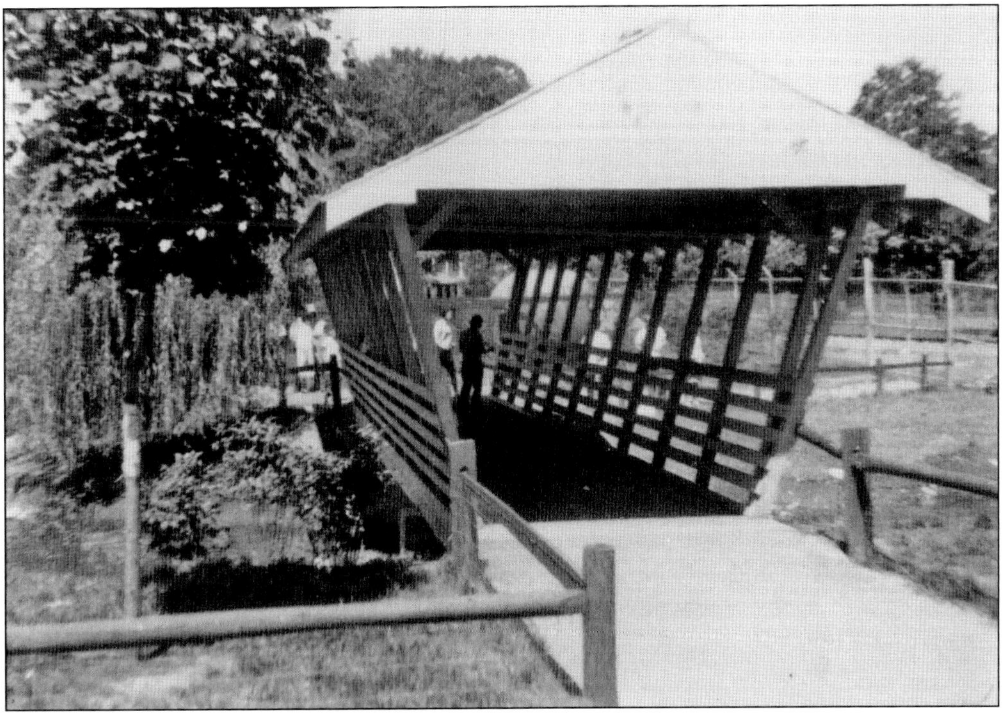

This covered footbridge is located in the Van Saun County Park in Paramus, Bergen County. It was built in the spring of 1963, has a length of 40 feet, and was set over the Van Saun Creek, which is a small tributary of the Hackensack River. The bridge is located near a small children's zoo at the park and welcomes visitors as they approach the zoo. The town of River Edge is also included within the boundaries of the park. Bergen County is one of a few counties in the state where no historic wooden covered bridges have been discovered and authenticated.

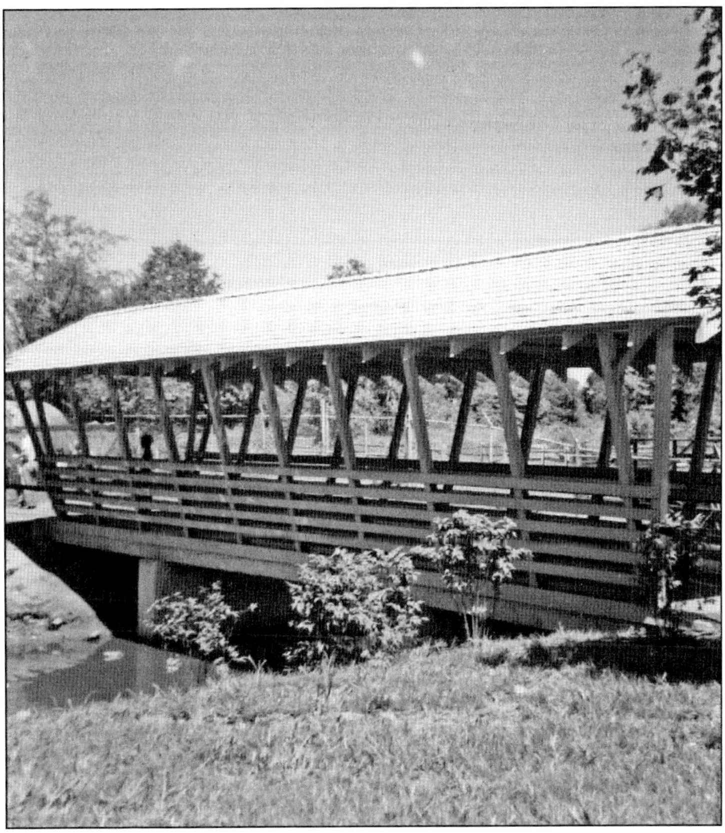

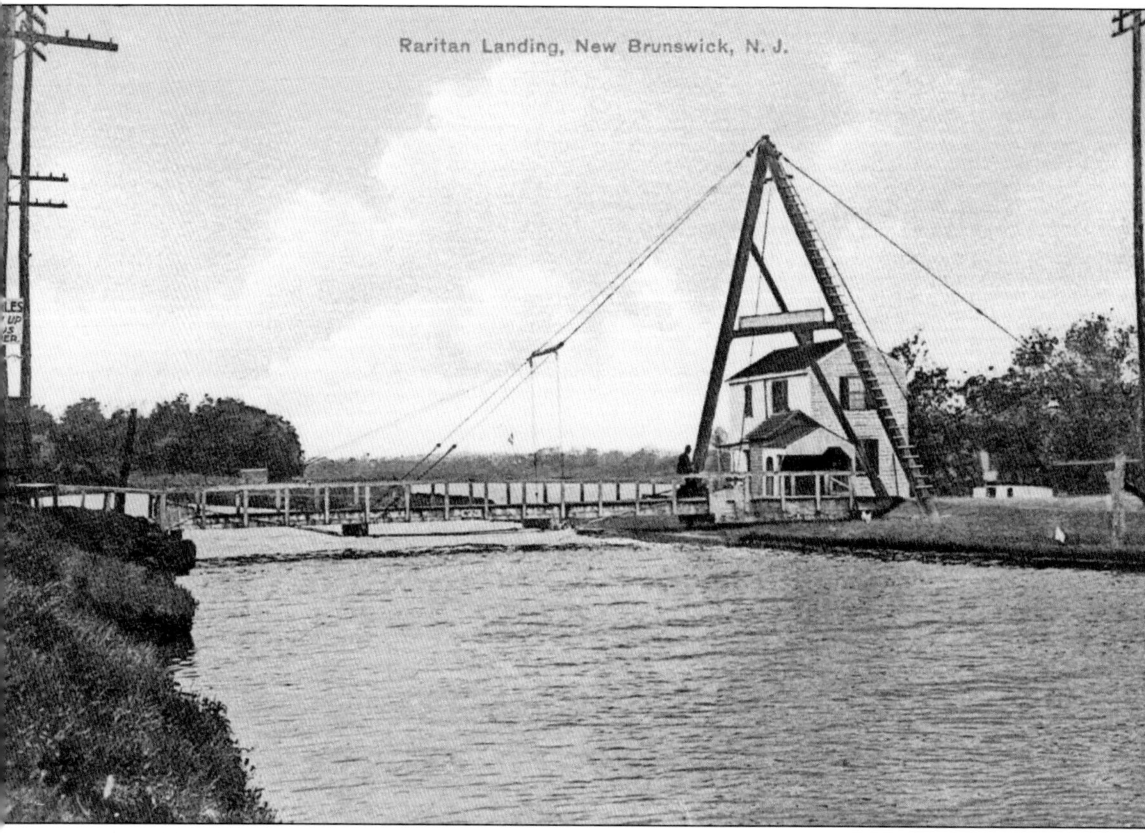

This old and unusual drawbridge marks the location of a once-thriving colonial river port known as Raritan Landing. In the early 1700s, a small community was established on the Raritan River, near New Brunswick. The spot was chosen because it was the farthest north that ships could travel upriver before it became too shallow to navigate. The landing quickly emerged as the center for local trade and served as a hub for imports and exports to and from the Raritan Valley. Agricultural goods and lumber brought to Raritan Landing from throughout central New Jersey were stored in warehouses, awaiting shipment to either New York or sometimes the Caribbean. Imported goods were off-loaded and taken by traders to stores and merchants throughout the area. During the 1770s, Raritan Landing was occupied by British troops and sustained damage during the Revolutionary War. After the war, the community rebuilt itself. Changes in transportation in the 1830s ended the need for Raritan Landing, and it slowly disappeared over the years. (Courtesy of the New Brunswick Public Library.)

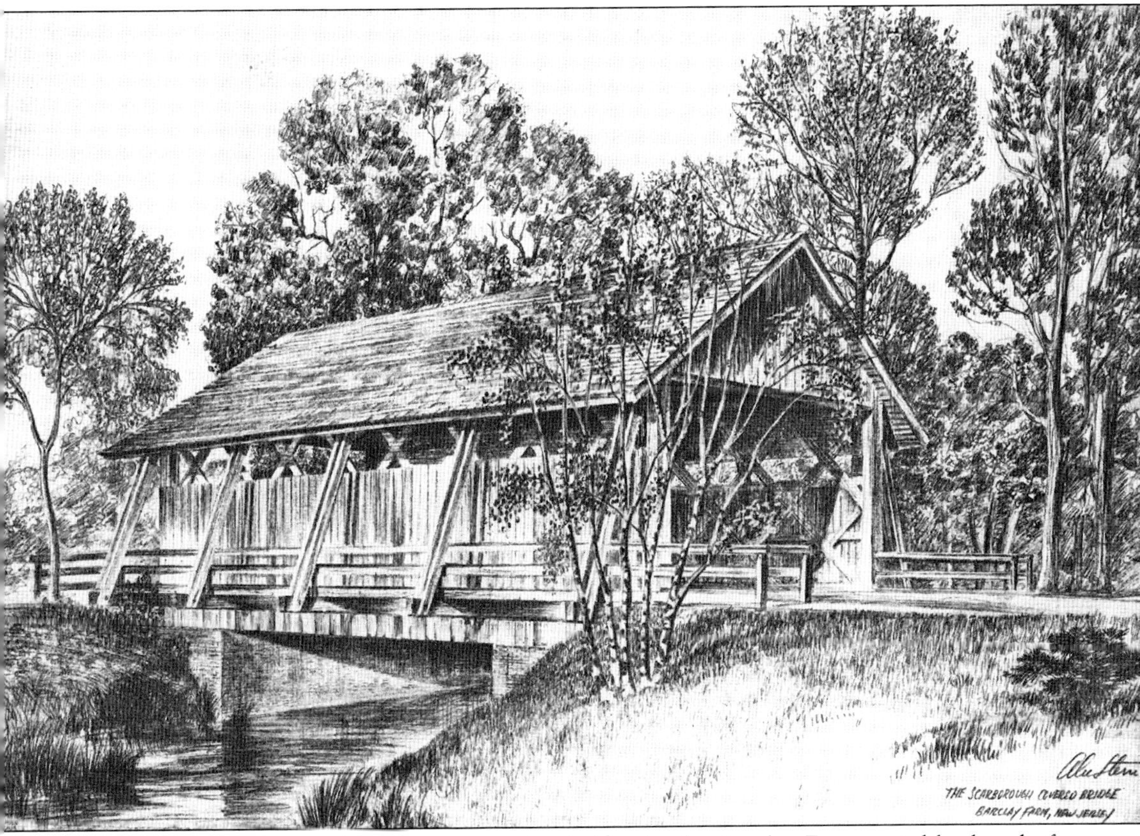

The Scarborough Covered Bridge is a wooden bridge in the Barclay Farms neighborhood of Cherry Hill, Camden County. It is a double-lane bridge for two lanes of traffic on Covered Bridge Road. There are also two sidewalks for bicyclists and pedestrians. Malcolm Wells designed the bridge, which was dedicated on Valentine's Day, February 14, 1959, with a reported 101 couples who kissed to confirm the romantic notion of a "kissing bridge." The bridge was named for Robert Scarborough, a developer of the Barclay Farms neighborhood, where a bridge was needed to extend the subdivision street system over the North Branch, a small tributary of the Cooper River. The decorative truss work within the bridge somewhat resembles a Town lattice but serves no function, as the bridge is a stringer type. The township recreated the original dedication ceremony for the 50th anniversary in 2009, which also included many couples who came to kiss inside the bridge.

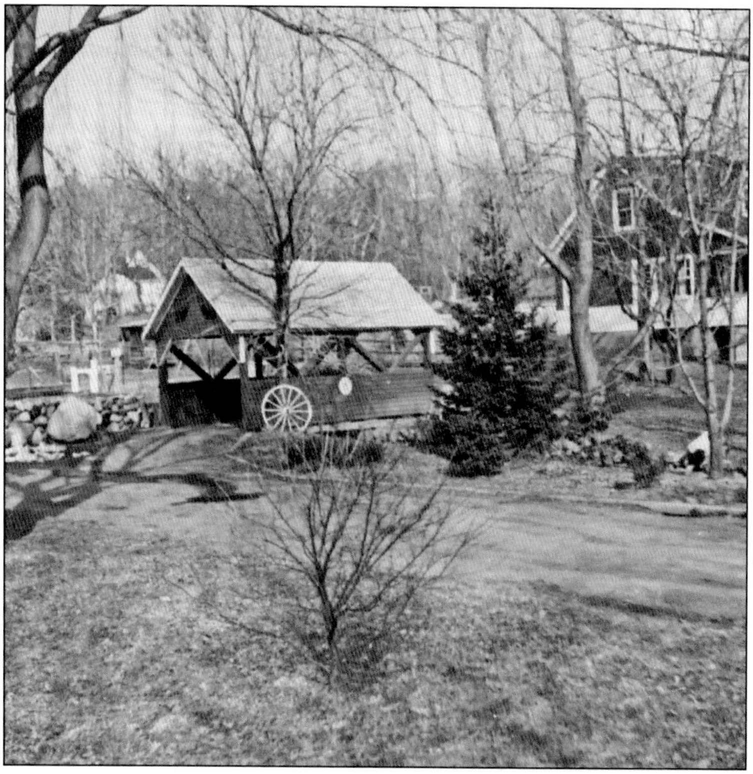

An old wagon wheel rests against the side of Ringer's Bridge. Daniel Ringer Jr. built the small bridge in 1966 on private property at the corner of Circle Avenue and Mountain Road in Franklin Lakes, Bergen County. Crossing a small tributary of the Ramapo River, the bridge is only 20 feet long and 12 feet wide.

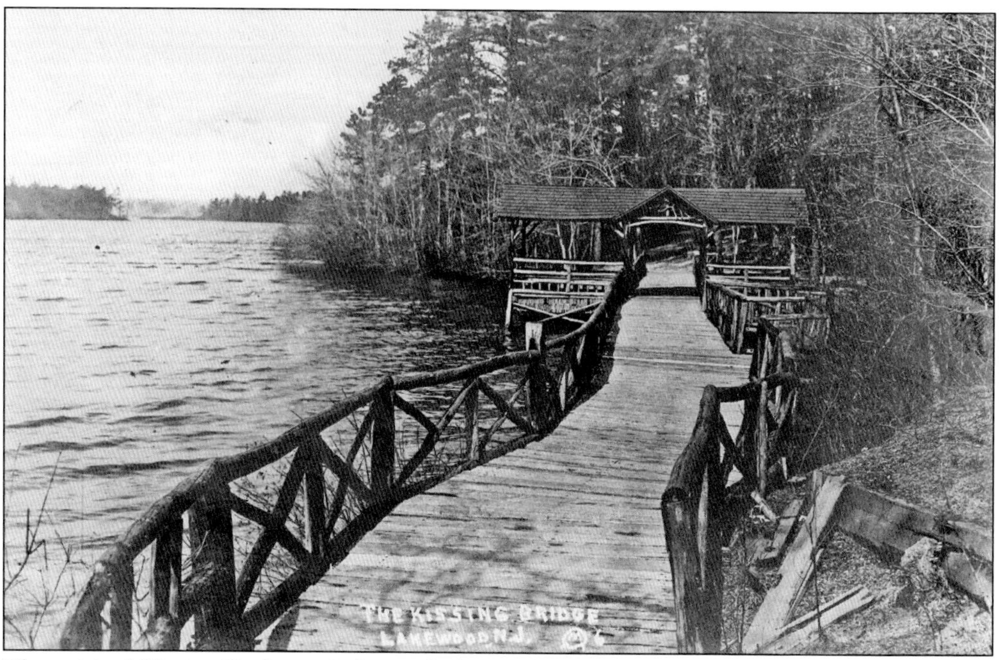

The original Kissing Bridge in Lakewood, Ocean County, spanned the sunken gardens of the great Gould estate. When Lakewood became the resort town of the rich and famous, this narrow log bridge could no longer handle the traffic. In 1905, a new Kissing Bridge was erected a short distance away, and a new arch bridge was also built.

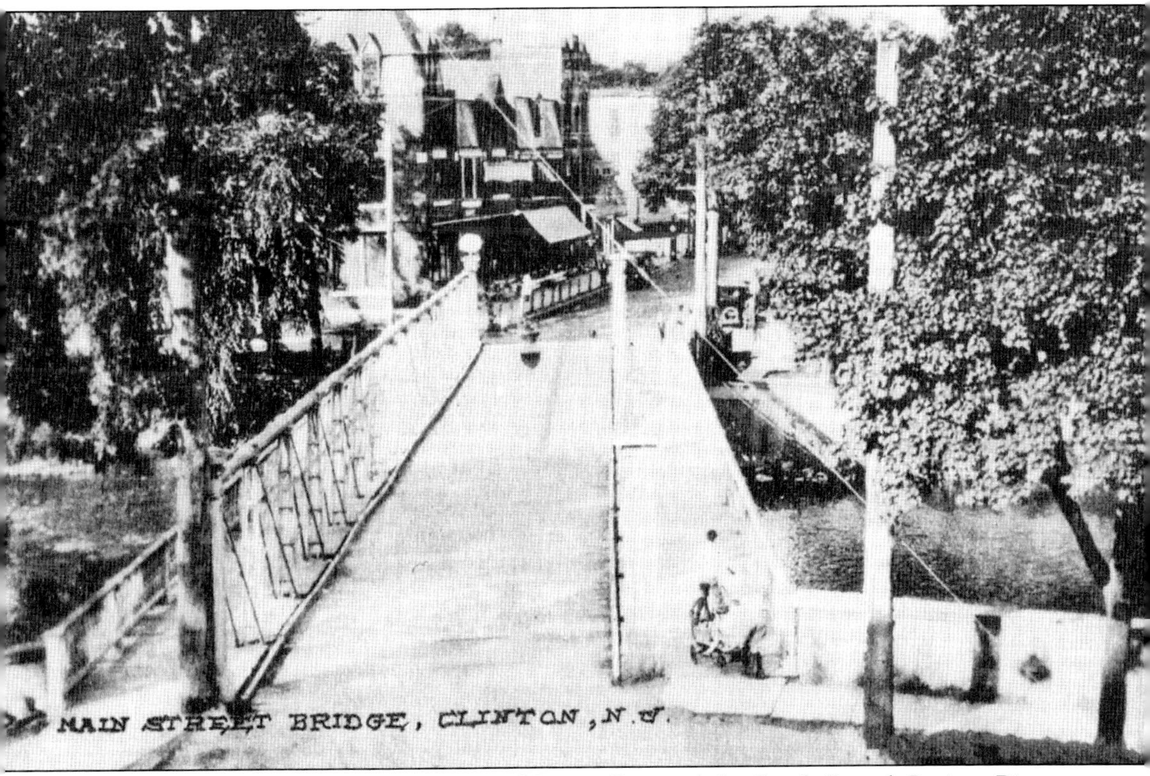

The first bridge spanning the confluence of Spruce Run and the South Branch Raritan River in Clinton, Hunterdon County, appears to have been a wooden covered bridge, according to research by the late town historian, Allie McGaheran. Prior to the bridge's construction, travelers had to ford the river, the crossing of which was rocky and difficult. A more recent cast iron bridge is shown. Due to the road and waterpower, the Hunt's Mill section grew in importance, outpacing the agrarian region and becoming a business and commercial center. With the mill changing hands several times, the residents sought a new name and decided to honor Dewitt Clinton, late governor of New York and developer of the Erie Canal. As written in the *Clinton Township History Essay* by Frank A Curcio, Clinton Township remained an essentially sleepy farm community until the mid-20th century, when drought brought on a project creating the Round Valley Reservoir and new roads made the township important again. (Courtesy of the Hunterdon County Historical Society.)

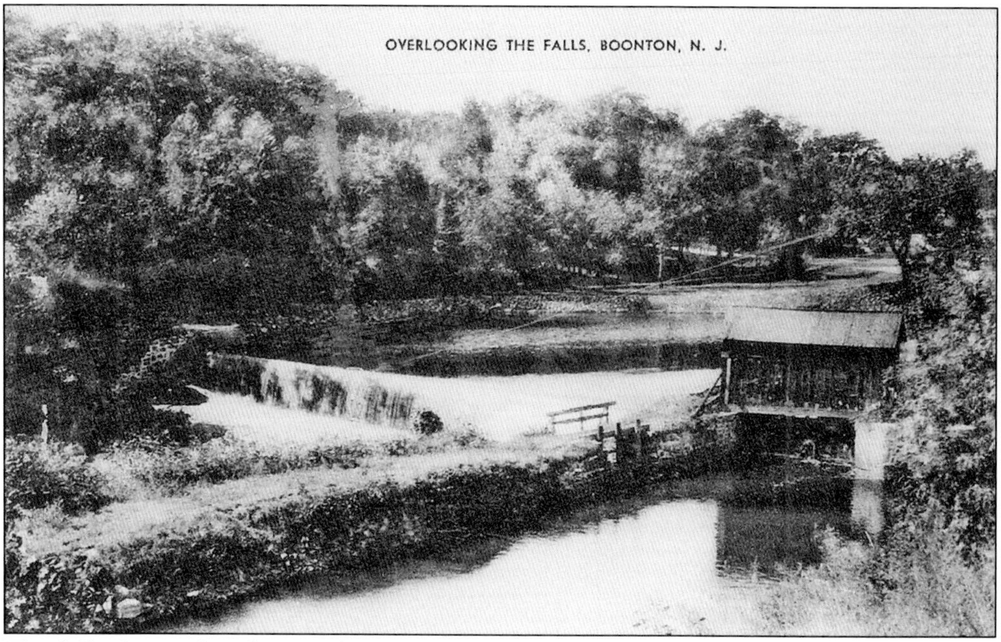

The difficulty in identifying and authenticating the existence of an old covered bridge in New Jersey is illustrated in these pictures from Boonton, Morris County. What appears to be a small covered bridge, perhaps a wooden stringer or kingpost structure, could not be verified from any other sources by the Boonton Holmes Library's local historian. In a book titled *Boonton Was an Iron Town*, a picture is included showing another section of the Morris Canal where a similar structure is identified as a gatehouse used to regulate the flow of water into the canal from the Rockaway River. Close up, the gatehouse below is similar in size but, clearly, is not a covered bridge. Consequently, the search and research must continue. (Above, author's collection; below, courtesy of the Boonton Holmes Library.)

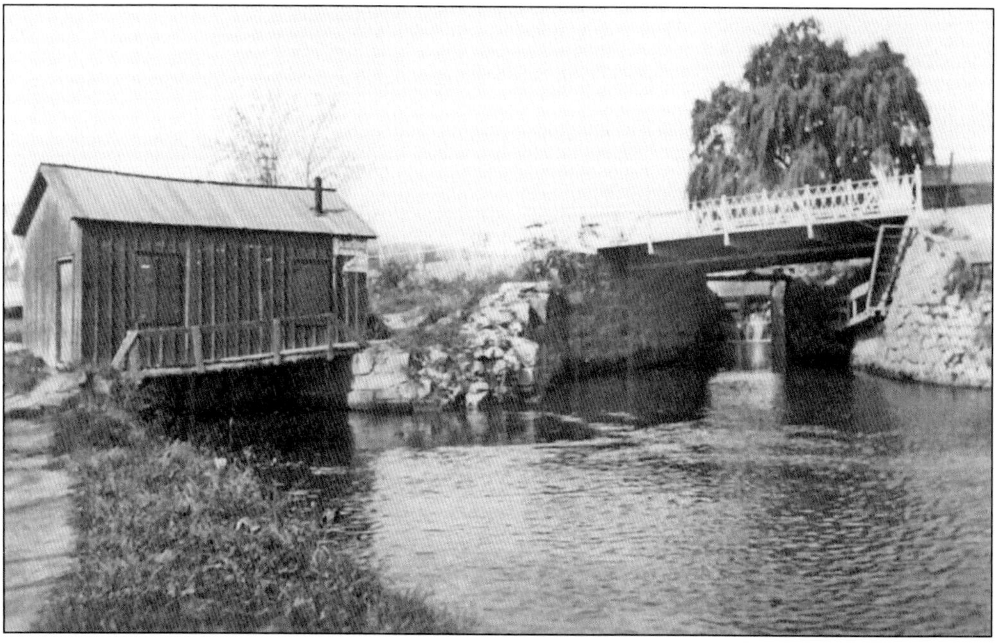

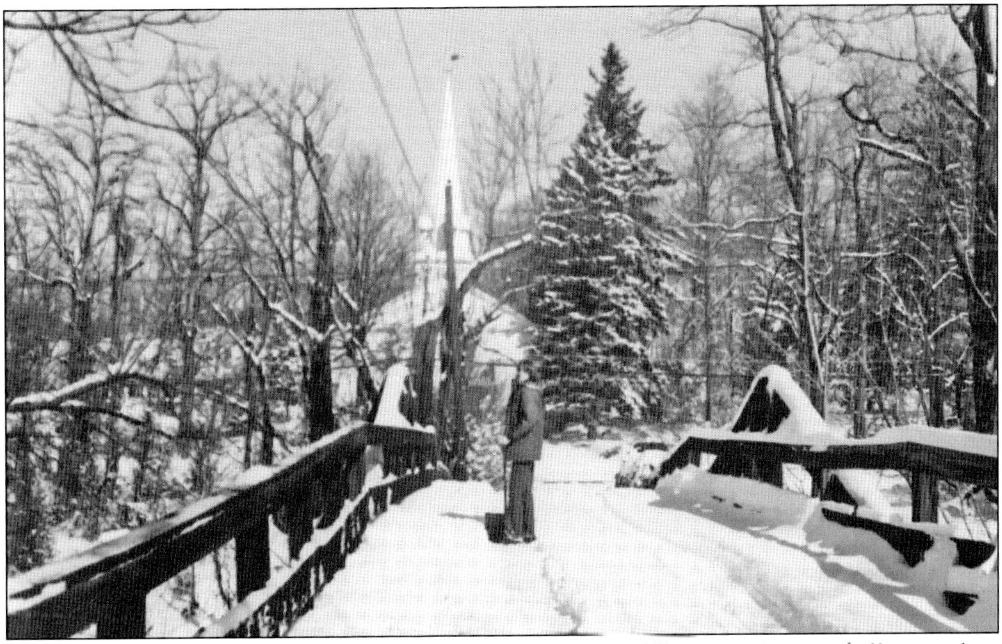

Another interesting story about a romantic shelter, built in Middletown, Monmouth County, does not live up to the preconceived notion of a quiet, covered hideaway for a pleasant evening stroll. This bridge was built over now electrified train tracks handling busy commuter traffic from the Jersey Shore to New York. The original bridge was an open wooden span, as seen above. The bridge connected private property, including a house built in 1739 and an apple orchard. Conrail built the new bridge to enable heavy emergency vehicles to cross and to close the sides. The covered bridge design was appropriate for this historic area, as seen with the old Reformed Church near the foot of the bridge. (Above, courtesy of Nancy O'Brien; below, author's collection.)

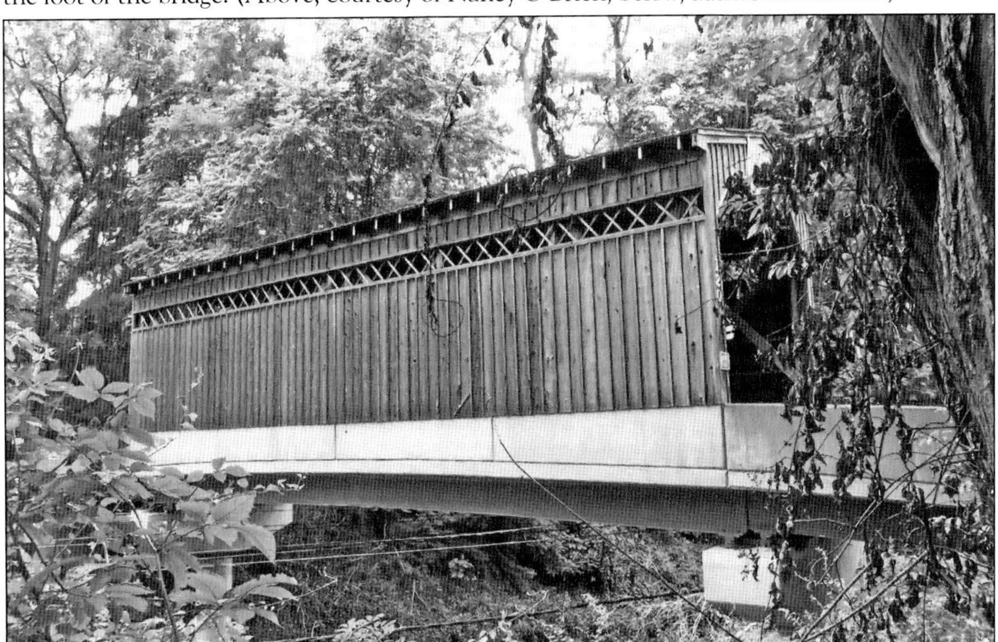

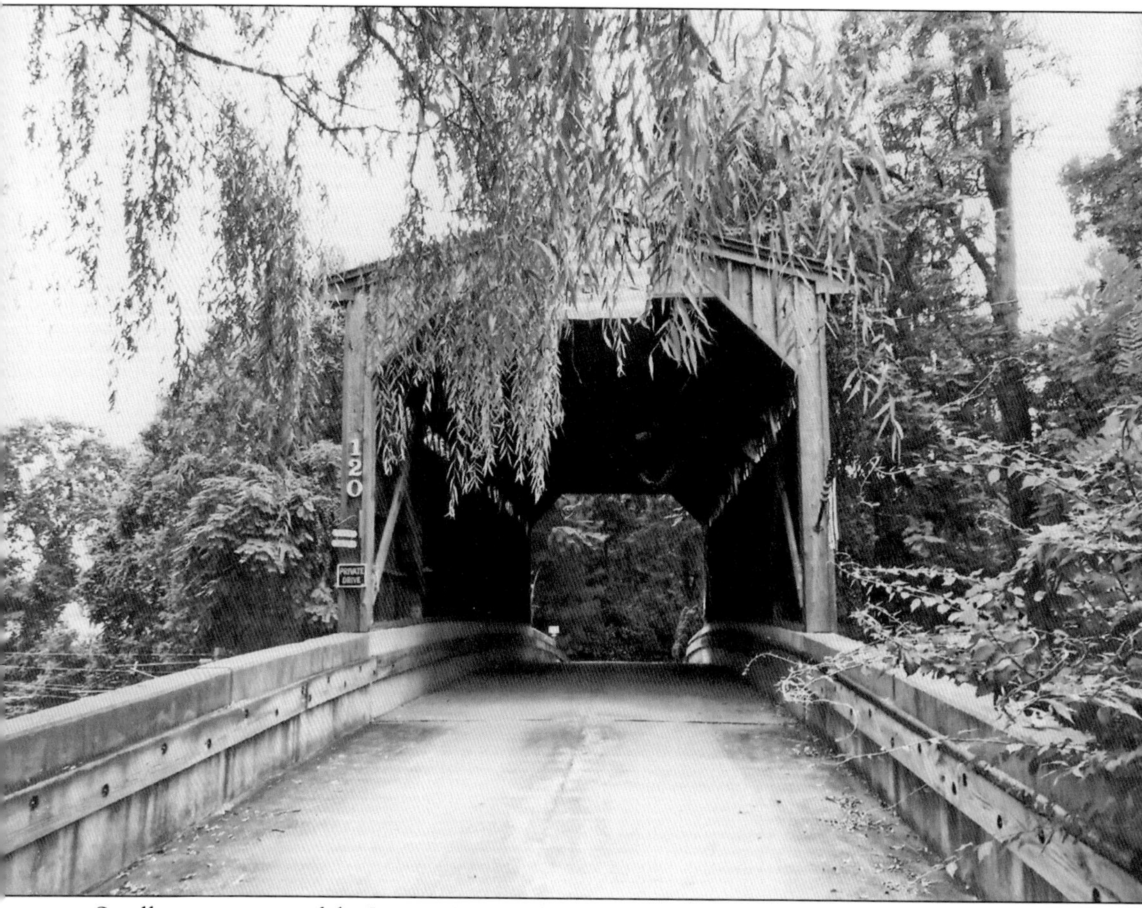

Small communities of the Lenape Navesink tribe were common throughout the area when the first known European landing in what would become Middletown Township occurred in 1609. Sea captain and explorer Henry Hudson anchored along the shores of Sandy Hook Bay and described the area as "a very good land to fall in with and a pleasant land to see." Shortly after the Dutch surrender of New Netherland to the British in 1664, a large tract of land known as the Navesink Patent, or Monmouth Tract, was granted to Quaker settlers. This land became the townships of Middletown and Shrewsbury. During the Revolutionary War, the British held much of Monmouth County, including Middletown, until the Battle of Monmouth. Upon the completion of a railroad junction in 1875, the town grew more rapidly, changing small and loosely connected fishing and agricultural villages into fast-growing suburbs.

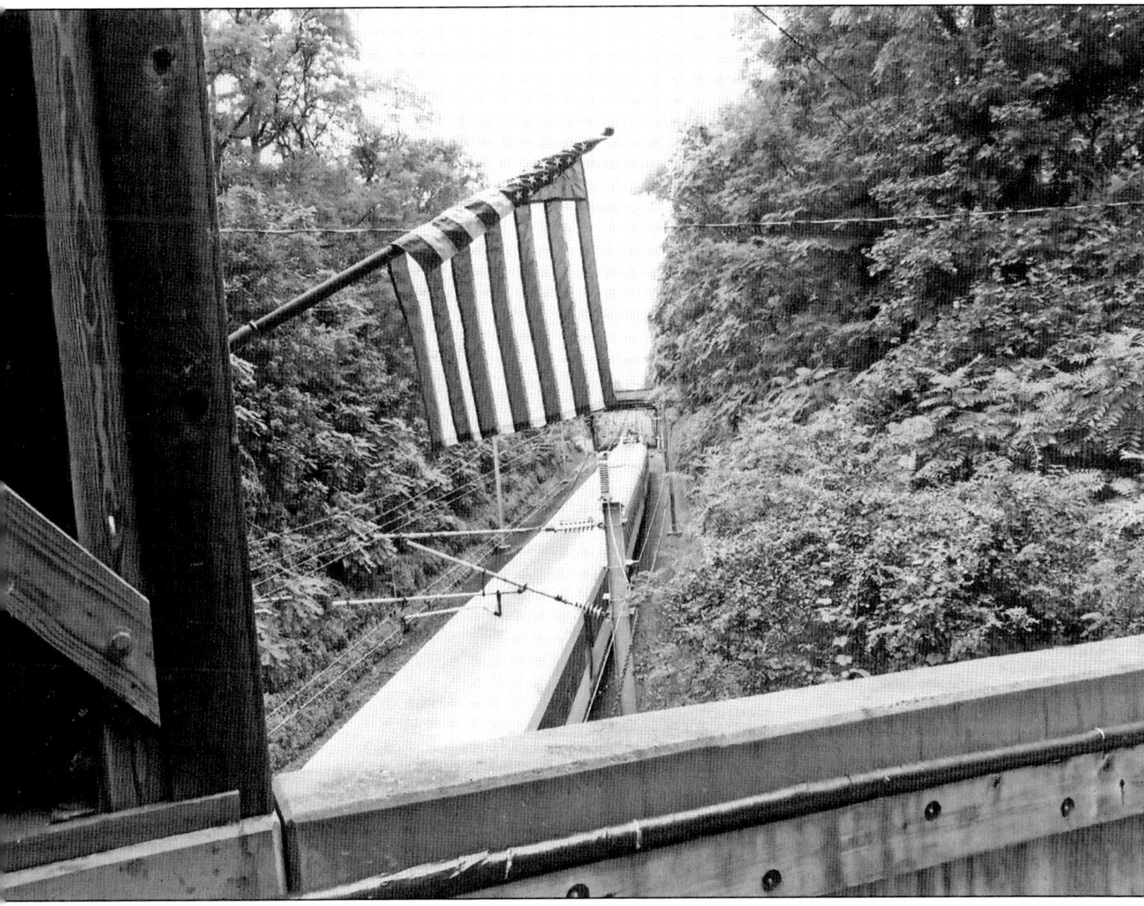

A New Jersey Transit Coast Line commuter train is seen passing under the Middletown Covered Bridge, heading north toward Newark and New York City. Conrail built this unusual bridge in 1980. The line was later electrified, with the wires running above the two tracks and under the bridge. According to "A Brief History of Conrail" found on the conrail.com website, Conrail began operations in April 1976, although its origins date back to the earliest days of railroading in North America. The oldest segment of what became Conrail was the Granite Railway Company, built in 1826 to carry granite blocks for the Bunker Hill Monument in West Quincy, Massachusetts. Six major railroad lines existed 150 years later, but in the early 1970s, all six had entered bankruptcy. The federal government responded by creating Conrail, the Consolidated Rail Corporation. In 1983, New Jersey Transit purchased the track and right-of-way for commuter operations.

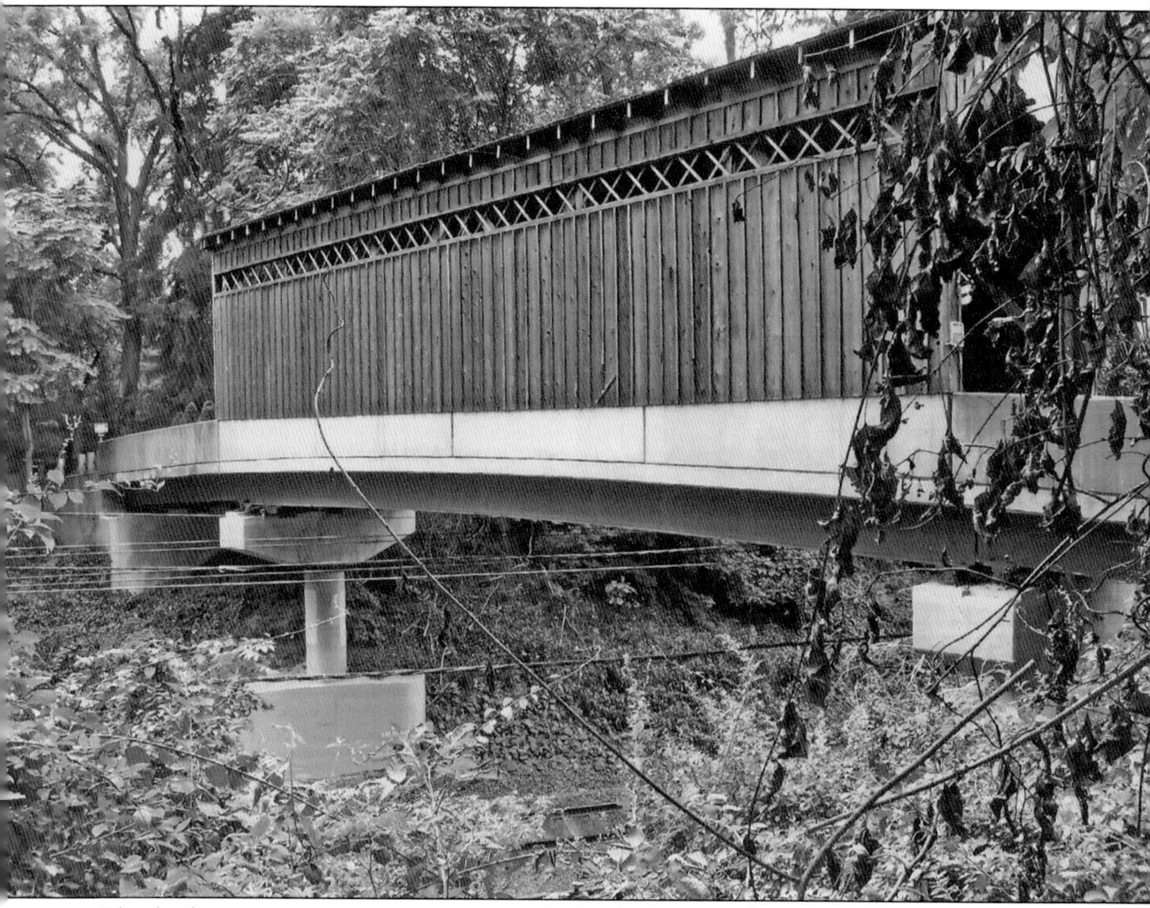

Clearly, the concrete stringer span, along with the abutments and two concrete piers straddling the railroad tracks, provide immediate verification that this is not an authentic covered bridge. The term generally used for this type of structure is a "romantic shelter," which refers to a nonauthentic covered bridge usually located on private property. Many romantic shelters are new, well-cared-for bridges built for nostalgic reasons and are intended to preserve memories or to enhance the historical ambiance of an area. They have been built as entrances to businesses or developments, as found in Cherry Hill, or to create a picturesque spot, as in Smithville. Others provide safe passage for children to cross a stream or even busy railroad tracks. Whatever the reason, these new covered bridges show an interest in preserving the memory of the old spans, which all too soon may be forgotten.

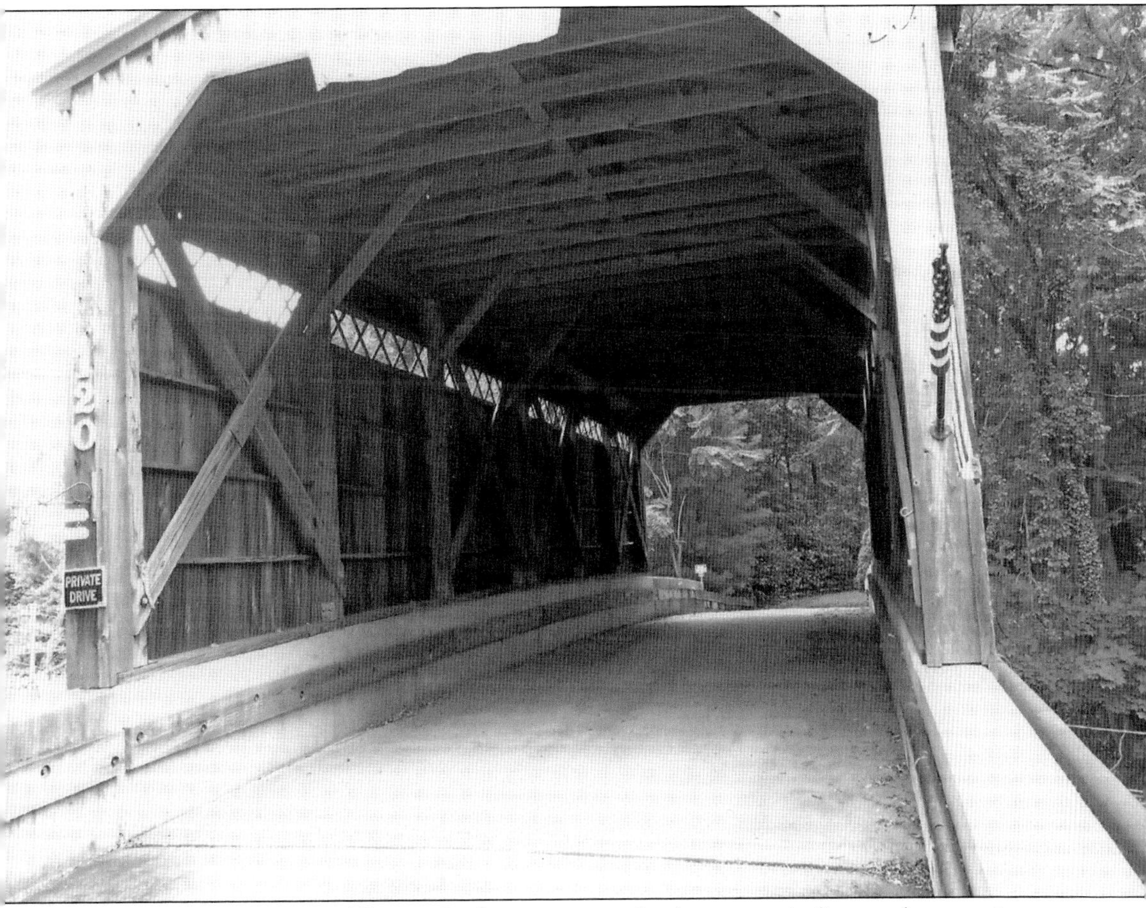

Using this picture of the Middletown Bridge as an example, the interior indicates what constitutes a romantic shelter. The wood crosspieces here obviously provide no support for this bridge; they are simply decorative. The same is true for the Scarborough Bridge (page 117). The Smithville Bridge contains original truss timbers from Pedricktown, but they no longer serve their original function (page 110). A very different conversation occurs when an authentic covered bridge is in need of restoration. Arnold M. Graton Jr. is known as a historical restoration builder of covered bridges, highly respected in his field of work. He has noticed a disturbing trend in covered bridge restoration: accepting state and federal money obligates a town to abide by standards regarding road clearance, load capacity, road width, and wind load. He writes, "We should be striving to preserve, not just the outward appearance of our covered bridges, but also the design, workmanship, and technology of the original builders."

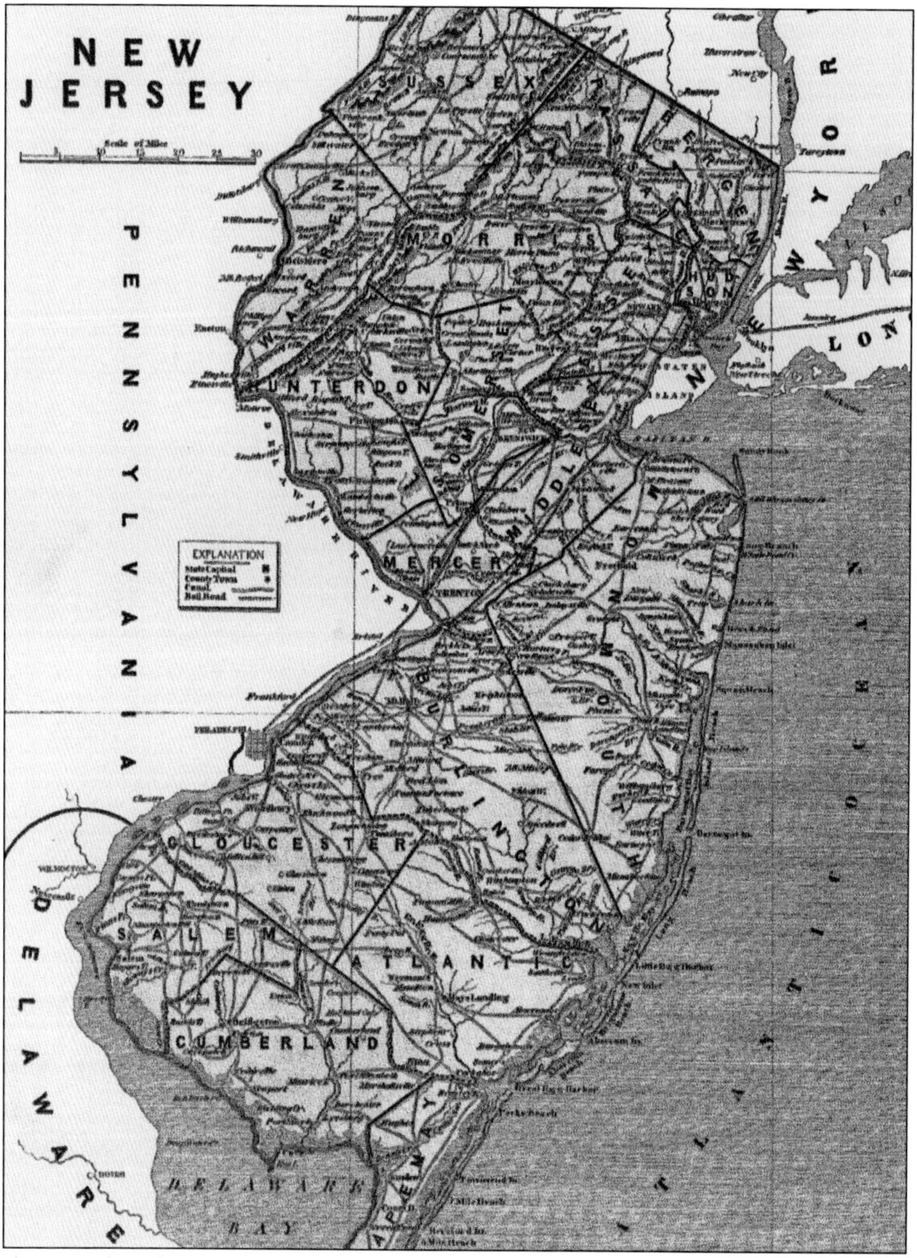

The first people to live on the land now known as New Jersey were the Delaware Indians. They lived here beginning at least 10,000 years ago. Their name means "original people" or "genuine people," and they spoke an Algonquian dialect. Later, small European trading colonies appeared where the towns of Hoboken and Jersey City are located today. The Dutch, Swedes, and Finns were the first European settlers in New Jersey. Looking at this 1865 map, the Delaware River Region is in the southwest (lower left) area of the state, the Southern Shore and Shore Regions are in the east and southeast (lower right and bottom), the Gateway Region is in the northeast (upper right) area, and the Skylands Region is found in the northwest (upper left) area, also with the Delaware River as the western border. New Jersey was named after the Isle of Jersey in the English Channel and became the third state on December 18, 1787.

BIBLIOGRAPHY

Abbott, Charles Conrad. *Wasteland Wanderings*. New York: Harper and Brothers, 1887.

Allen, Richard Sanders. *Covered Bridges of the Northeast, Rev. Ed.* Brattleboro, VT: The Stephen Greene Press, 1974.

Barber, John W. and Henry Howe. *Historical Collections of the State of New Jersey*. New York: S. Tuttle Company, 1844.

Beitel, Herbert. *Cape May County: A Pictorial History*. Virginia Beach, VA: Donning Company, 1988.

Borven, F.W. *History of Port Elizabeth, Cumberland County, NJ*. Philadelphia: Lippincott Company, 1885.

Boyer, Charles, and Philip Cohen, ed. *Rambles Through Old Highways and Byways of West Jersey*. Camden, NJ: Camden County Historical Society, 2003.

Brydon, Norman F. *Of Time, Fire and the River*. New Vernon, NJ: New Vernon Business Service, 1970.

Dale, Frank T. *Bridges Over the Delaware*. New Brunswick, NJ: Rutgers University Press, 2003.

Egee, Wilmer, and Edmund Burk. *Swedesboro Yesterday and Today*. Philadelphia: The Literary Bureau, 1910.

Henn, William F. *The Story of the River Road*. Published by William F. Henn, 1978.

Hood, John. *Index of Colonial and State Laws, 1663–1903 Inclusive*. Camden, NJ: Sinnickson Chew and Sons Company, 1905.

Johnson, Paul. *"Art" and the Language of Progress in Early—Industrial Paterson: Sam Patch at Clinton Bridge*. American Quarterly 40:4 (1988), The American Studies Association, Johns Hopkins University Press.

Lossing, Benson J. *Pictorial Field Book of the Revolution, Volume II*. New York: Harper and Brothers, 1851.

McMahon, William. *Historic Town of Smithville*. Egg Harbor City, NJ: The Laureate Press, 1967.

Wendt, Peter C. *Boonton Was an Iron Town*. Morristown, NJ: Compton Press, Inc., 1976.

Discover Thousands of Local History Books Featuring Millions of Vintage Images

Arcadia Publishing, the leading local history publisher in the United States, is committed to making history accessible and meaningful through publishing books that celebrate and preserve the heritage of America's people and places.

Find more books like this at
www.arcadiapublishing.com

Search for your hometown history, your old stomping grounds, and even your favorite sports team.

Consistent with our mission to preserve history on a local level, this book was printed in South Carolina on American-made paper and manufactured entirely in the United States. Products carrying the accredited Forest Stewardship Council (FSC) label are printed on 100 percent FSC-certified paper.

MADE IN THE USA